PRODUCING PREDATORS

"*Producing Predators* excels in its positioning of work as a useful tool to understand the complex contours of human-animal relations in the nineteenth-century West."
—Karen Jones, *History: Journal of the Historical Association*

"This book will be of great interest to environmental historians, historical geographers, and scholars who examine human-animal relations through time. . . . Much more than simply another book on wolves, Wise offers an analysis which is part environmental history and part account of the specificities of colonialism in the Rockies."
—Stephanie Rutherford, *H-Environment*

"*Producing Predators* should be on the short list for scholars interested in further exploring the tensions embedded in Western history, particularly the convoluted intersections between labor, race, and ecology that Wise so deftly uncovers in his first book."
—Frank Van Nuys, *Annals of Wyoming*

"*Producing Predators* is an interesting, well-written, and informative account of the Northern Rockies ecosystem. . . . Specialists will find it a well-executed study of colonialism in the American West."
—Adam Sowards, director of the Institute for Pacific Northwest Studies at University of Idaho

"An academic historian, Wise is especially insightful in his descriptions of the ways that environmental practices gain cultural legitimacy as aspects of larger ideologies about assimilation, economics, and labor."
—Sarah E. McFarland, *American Indian Quarterly*

PRODUCING PREDATORS

*Wolves, Work, and Conquest
in the Northern Rockies*

MICHAEL D. WISE

University of Nebraska Press | Lincoln

© 2016 by the Board of Regents of
the University of Nebraska

Part of chapter 2 originally appeared in "Killing Montana's Wolves: Stockgrowers, Bounty Bills, and the Uncertain Distinction between Predator and Producer," *Montana: The Magazine of Western History* 63 (Winter 2013): 51–67. Part of chapter 3 was originally published in "Colonial Beef and the Blackfeet Reservation Slaughterhouse, 1879–1895," *Radical History Review* 110: 59–92. Copyright 2011 MARHO: The Radical Historians Organization, Inc. All rights reserved. Republished by permission of the copyright holder and the present publisher, Duke University Press, www.dukepress.edu.

All rights reserved

Library of Congress Cataloging-in-Publication Data
Names: Wise, Michael D.
Title: Producing predators: wolves, work, and conquest in the northern Rockies / Michael D. Wise.
Description: Lincoln: University of Nebraska Press, 2016. | Includes bibliographical references and index.
Identifiers: LCCN 2015042852 | ISBN 9780803249813 (cloth: alk. paper) | ISBN 9781496222336 (paperback) | ISBN 9780803290464 (epub) | ISBN 9780803290471 (mobi) | ISBN 9780803290488 (pdf)
Subjects: LCSH: Wolves—Rocky Mountains—History. | Wolves—Control—Rocky Mountains—History. | Human-animal relationships—Rocky Mountains—History. | Ranching—Rocky Mountains—History. | Rocky Mountains—History.
Classification: LCC QL737.C22 W574 2016 | DDC 599.7730978—dc23 LC record available at https://lccn.loc.gov/2015042852

Set in Ehrhardt by M. Scheer.
Designed by N. Putens.

For Jen and Augie

Contents

List of Illustrations	ix
Acknowledgments	xi
Introduction	xv
1. Wolves and Whiskey	1
2. Beasts of Bounty	23
3. Making Meat	47
4. The Place That Feeds You	69
5. Unnatural Hunger	99
Conclusion	133
Notes	139
Bibliography	161
Index	175

Illustrations

1. Charles Marion Russell,
   Wolves at the Wagon-Train   5

2. Reaching for entrails at the
   agency slaughterhouse   62

3. *Agency Slaughterhouse, 1905*   63

4. White Grass Chapter PFLA, ca. 1924   94

5. *Five Minutes' Work*   116

6. Work of the unnaturalized
   Basque sheepherder   121

7. Jetta Hamilton Grey with wolf pup   136

8. Janet Hogmire with coyote   137

Acknowledgments

It is a delightful and selfish privilege to publicly thank all of the people and institutions who have supported this book—my first one—with their support and criticism over the years. So my initial expression of gratitude goes to this book's readers, especially the ones who bristle at the aggrandizing tone of indulgent acknowledgment sections. Thank you for suffering through the following paragraphs.

I first envisioned this book at an environmental history workshop hosted by Montana State University and the University of Wisconsin at Chico Hot Springs, near the Yellowstone River, in the winter of 2007. We went wolf watching in Yellowstone National Park the afternoon before I presented my paper, and I think that experience helped inspire the attention of the out-of-town scholars who may otherwise have been uninterested in my research about the history of the nineteenth-century wolf pelt trade. These visitors included Bill Cronon, Nancy Langston, Don Mitchell, Gregg Mitman, and Bob Wilson, and they all gave me generous and provocative feedback that day, as did my many mentors at Montana State—Rob Campbell, Tim LeCain, Mary Murphy, Sara Pritchard, Michael Reidy, Billy G. Smith, and Brett Walker. It is no stretch to say that this project benefited at an early stage from the input of many brilliant and respected scholars, and for that I am grateful.

This book continued its development as a dissertation guided by my coadvisors at the University of Minnesota, David Chang and Susan Jones, who challenged me to read deeply in the fields of Native history, political economy, and the history of animal-human relationships in order to cobble together what seemed at first to be an unusual approach to studying the history of wolves and cattle in the North American West. I would also like to thank my other mentors at the U, George Henderson, Stuart McLean, Jeani O'Brien, and Jeffrey Pilcher, for their criticism, encouragement, and advice.

Over the years, I also had the good fortune to present drafts of early chapters to the University of Minnesota's American Indian Studies Workshop and at national meetings of the American Society for Environmental History, the Native American and Indigenous Studies Association, and the Western Historical Association. I owe significant debts to four organizations that funded my research and to archival professionals who assisted me with the challenges of locating collections and manuscripts. The Cody Institute for Western American Studies, the Denver Public Library, the Newberry Library, and the Social Science Research Council each provided generous fellowships to pay for archival research, and a long list of individuals helped me to conduct it: Robert Olley at the Bronx Zoo Library; Kurt Graham, Diana Jensen, Bob Pickering, and Mary Robinson at the Buffalo Bill Historical Center; Wendel Cox, Claudia Jensen, and Ellen Zazzarino at the Denver Public Library; Patricia Molesky at the Glenbow Museum; Kim Scott at the Montana State University Special Collections; Andrew Knight Jr. at the National Archives and Records Administration; John Aubrey and Diane Dillon at the Newberry Library; and Jana Wilson at the Stockmen's Memorial Foundation. A special thanks goes to the staff at the Montana Historical Society, especially Rich Aarstad, Jeff Malcomson, and Brian Shovers.

A number of other scholars and editors, some friends and some strangers, offered their advice, support, and criticism. Bridget Barry believed in this project nearly from the beginning, and I would like to thank her and the rest of the staff at the University of Nebraska Press. I would also like to thank Jaime Allison, Etienne Benson, Michael Bottoms, Janet Browne, Jon Coleman, Boyd Cothran, Chris Crawford, Tracey Deutsch, Melanie

DuPuis, Colter Ellis, Bill Farr, Steven Fountain, Rob Gilmer, Joe Haker, Molly Holz, Andrew Isenberg, Karl Jacoby, Jerry Jessee, Regina Kunzel, Rosalyn LaPier, Christine Manganaro, Saje Mathieu, Paul McCutcheon, James McWilliams, Nic Mink, Dave Morton, Kevin Murphy, Andy Paul, Trey Proctor, Megan Raby, Harriet Ritvo, Ryan Shapiro, Adam Sowards, Rebecca Woods, and this book's anonymous reviewers for their advice and critical feedback.

My colleagues at the University of North Texas and other nearby institutions welcomed me to Denton four years ago and have supported this project with their interest and friendship. Laila Amine, Agatha Beins, Charles Bittner, Sandra Mendiola, Chad Pearson, Clark Pomerleau, and Priscilla Ybarra all read portions of the manuscript and offered their critical perspectives. I also benefited from thoughtful conversations with Neilesh Bose, Sophie Burton, Mike Campbell, Guy Chet, Andrew Graybill, Harland Hagler, Rick McCaslin, Mick Miller, Rachel Moran, Marilyn Morris, Todd Moye, Walt Roberts, Gus Seligmann, F. Todd Smith, Sherry Smith, Sonny Solis, Nancy Stockdale, Michael Thompson, Andrew Torget, Liz Turner, and Kelly Wisecup. In addition to sharing their wisdom, Baird Callicott and Katherine and Hank Eaton also put their roofs over my head.

A number of friends and family members deserve recognition for living with this book over the span it has taken to create it. Stefanie Bergh shouldered the weight of this project for much of its life. Tuba Adam, Mike Benno, Isaac King, Steve Schwaeber, and Glenn Tolle all listened patiently about wolves over thousands of miles that we pedaled together. Andy Fischer and his family frequently hosted me in Montana on research trips after I moved away. And friendships with a number of nonhuman companions also animated and enriched my work. A furry thanks to the late Axel, Bingley, Darcy, the late Mrs. Collins, and Stella.

My grandparents, Walt and Betty Lienemann and Leonard and Anna Marie Wise, fanned my curiosity about the past from an early age. I hope this is a book that Leonard and Walt would have enjoyed reading. My parents, Don and Debbie Wise, stuck with me through some tumultuous changes that occurred in the final year of publishing this book. I am grateful for them, and for my brother, Brian, for their unconditional love and support.

Finally, I owe an enormous debt to my partner, Jennifer Jensen Wallach, whose love and friendship carried this book as it hit the homestretch during a dramatic moment in our lives. She is the most kind and exciting thinker I have ever known, and I feel lucky to be with her each day. The strengths of the following pages owe much to her inspiring warmth, generosity, and confidence, and so it is to Jen, and to our son, Augie, that I dedicate this book with all of my love.

Introduction

LIVING LIKE A WOLF

In January 1870 a young immigrant named Peter Koch used the metaphor of "living like a wolf" to express the realities of his new life in the frontcountry of the Northern Rockies. For seventy-five dollars a month, Koch labored as a clerk at the confluence of the Musselshell and Missouri Rivers, selling whiskey, strychnine, and other provisions out of a small storehouse to men who went "wolfing" on the plains above the river bluffs, killing wolves, coyotes, and other carnivores by baiting them with poisoned animal carcasses. Koch reflected on the predatory characteristics of this work in a series of letters written home to his family in Mississippi. In contrast to the bounty payments offered for killing wolves at his boyhood farm, which he had left a year earlier, Koch remarked that "Montana is not civilized enough for putting a prize on their scalps, but their skin is worth $2.00 here and probably more in the states." Instead of killing wolves for bounty, Koch and his employer, George Clendinnen, took advantage of an eastern demand for furs by shipping hundreds of wolf pelts to wholesalers in St. Louis and beyond and outfitting local wolfers with poisons and other supplies. Like the emerging class of sharecroppers back in the Reconstruction South that Koch had left behind, these wolfers risked insolvency with each harvest, "in debt so deeply," he recalled, that "it takes half the winter to get clear." Observing his own complicity in this system of indenturing human labor

xv

to kill and transform animal bodies into property, Koch admitted that "a man has to live pretty much like a wolf, if he's in the business."[1]

Koch mastered this business over the following decades, maturing to wealth and prominence as one of the region's major financiers and political figures. As a principal of the Bozeman National Bank, he managed networks of capital that profited from the expansion of cattle ranching and the industrial extraction of animal flesh. As the founding treasurer of Montana State University, originally dubbed the Agriculture College of the State of Montana, he pressed science and the study of the nonhuman world into the service of meat production and capital accumulation. He accomplished all of this while remaining a likeable member of his community, respected for his philanthropy and civility, "living like a wolf," perhaps, only in his own youthful estimation.[2] But his career benefited from broader environmental and cultural changes that seated him at the top of a new colonial food chain. By the early twentieth century, the livestock industry had overrun the Northern Rockies, incorporating what became parts of Montana and Alberta into a garden of meat for consumers across the globe.[3] This colonial conquest and privatization of land and animals redirected ecological flows of work and energy from the sun and the grass into the muscles of cattle sold for slaughter.[4] In that sense, Koch, along with stock growers and other stakeholders in the beef industry, replaced wolves as the Northern Rockies' apex predators.[5]

The more completely that cattlemen achieved this transformation, however, the more they sought to disavow their own predatory inclinations and to dissociate themselves from the wolves that preceded them. This dynamic pervaded social relationships throughout the conquest and colonization of the frontcountry. It also motivated campaigns to exterminate gray wolves and other wild carnivores from the Northern Rockies in the late nineteenth and early twentieth centuries, crusades that cost stock growers millions of dollars more than they saved. As this book demonstrates, wolf eradication served its primary purpose not as an economic fix for industrial agriculture but as part of a broader set of environmental and cultural practices that established new social boundaries between predators and producers. These boundaries facilitated the transfer of wealth to regional capitalists like Koch by representing stock raising and wage labor as productive forms of work in opposition to precolonial alternatives such as hunting and subsistence foraging. In many cases, the language of the colonial apparatus explicitly reflected this new boundary between predation and production. Wolves and wild carnivores were "non-producers" (as officials of the U.S. Bureau of Biological Survey put it), predators whose existence directly contested the work of the region's most significant economic actors, livestock growers.[6] But it would be a mistake to focus this history of predator-prey relationships on words alone, since in a day's work of killing, physical practice often exceeded the communicative significance of language. The dissociation of predation and production was not so much spoken about or recorded in text as it was performed through the daily tasks of carving out livelihoods from the bodies of animals, a social distinction felt and sensed as well as imagined.[7]

These activities drew non-Native settlers into contact with the many indigenous communities of the frontcountry. The Blackfoot men and women who reluctantly shared their homelands with Koch and his fellow newcomers linked their own productive labor as meat eaters with the predatory lives of wolves and other nonhuman carnivores.[8] They maintained that they learned to hunt bison through an empathetic relationship with wolves—they followed and observed wolves and tried to insert themselves as wolves into existing bison-wolf interactions. They performed wolf in their ceremonies and in the field. In reciprocal exchange, they left meat at their kill sites as offerings for wolves and other carnivores.[9] The Blackfoot developed sophisticated modes of kinship with wolves that transcended colonial understandings of the separation between human and animal. For the most part, these practices were entirely misapprehended by Indian Office personnel. Even experienced Indian agents commented on the Blackfeet's "wolfishness" with incredulity. Remarking on their practices of slaughtering cattle by running and shooting them in an open field, one official in 1885 claimed that the Blackfeet were "nearer to barbarians than anything I have ever seen."[10] This conclusion misunderstood the fact, however, that for the Blackfoot, turning cattle into prey by hunting them was an ethical means of transforming an animal's body into food. The Anglo-American tradition of the birth-to-death domestication of meat animals was not only an unfamiliar custom but also an unnecessary one within a world of animal-human relations structured around kinship rather than ownership. Rather than yield to the fiction of humane slaughter, the Blackfoot participated directly in the predatory labors of the animal world, a world in which they included themselves as animals.[11]

Along with their campaigns to eradicate wolves from the Northern Rockies, stockmen and colonial authorities in the United States and Canada also sought to purge Blackfoot and Salish-Kootenai communities of their own so-called predatory cultures through campaigns of assimilation that circumscribed hunting and reordered indigenous patterns of work, food, and livelihood. Dissociating predation from production simplified the moral complexities of colonization and conquest. It allowed non-Native ranchers and landowners to justify their dispossession of indigenous communities by claiming that their labor had rescued the land itself, that they had transformed the Montana-Alberta borderland into a productive pasture from a predatory wilderness.[12]

The Blackfoot, Salish, Kootenai, and other indigenous people in the region, however, were not simply shunted aside by this inexorable wave of colonialism. Despite suffering from heavy tolls of disease, theft, and outright murder as American and Canadian newcomers appropriated their homes and resources, their communities necessarily adapted to cattle and private property. They became participants, willingly and unwillingly, in a regional political ecology structured by the separation of predators from producers. Federal grazing leases and annuity programs, along with the practical extermination of bison and other wild game animals, drew many Native families into cattle ranching. Boarding schools and other organs of cultural assimilation taught children to give up hunting for baking and to earn their bread indirectly by selling their work for wages. The forced privatization of communally owned reservation lands under the allotment acts further compelled these transformations. Native people would no longer live like wolves off their tribal land; they would instead produce income from individually owned farms, subject to Anglo-American standards of labor and land tenure.[13]

Indigenous resistance to these changes existed mostly in the form of reshaping specific political and economic prerogatives of the borderland's colonial transformation to serve and maintain their own precolonial social obligations. The creation of public bounty payments to support wolf eradication, for instance, not only encouraged the Northern Rockies' transformation into cattle country by destroying wild carnivores but also provided a new opportunity for unwaged labor that was continuous, in some respects, with the hunting and trapping opportunities of the nineteenth century's intercultural fur trade. Opposition to the allotment of the Blackfeet Reservation led to the creation of an agricultural cooperative that coordinated the use of individual allotments for the broader social project of Blackfeet food provisioning. Although, in both these cases, the historical outcomes did not represent a return to past experiences, the initiatives nevertheless drew from precolonial conceptions of the social and environmental links between work and livelihood that sought to connect, rather than dissociate, the ecological labors that united human and nonhuman animals. A community's economic value did not simply materialize through the sum total of the individuated, productive labor of its discrete members; it coursed instead through holistic systems that connected human and nonhuman beings with the lands they inhabited through mutual cycles of predation and production. Blackfoot scholar Betty Bastien has described the unraveling of these "kinship alliances" as the fundamental process of colonialism in Alberta.[14] Her crucial observation explains much about the region's history, but a closer analysis of Native and non-Native labor, resistance, and adaptation during the conquest and privatization of the Northern Rockies region shows that the colonial process was left unfinished. Predators still produced—wolves, unwaged hunters, and others continued to generate and distribute value through formal and informal social and ecological networks, despite denials of their economic agency as nonproducers. Likewise, cattle producers sustained themselves not only through their own individual labors but through the broader social and ecological metabolisms that transformed their cattle into money. Despite colonial disavowals of the connections between predation and production, those relationships persisted.

Conquering the frontcountry for industrial meat production required more than just taking possession over Native land and labor. It also required ranchers and colonial authorities to redefine what constituted legitimate forms of work within a wide range of cultural, environmental, and political practices that distorted predators as a class of violent and dangerous nonproducers—as nonworkers. In this regard, the colonization of the Northern Rockies occurred alongside a broader sweep of global capitalism that created a world of inequalities where the working poor did lots of work and the idle rich did little work. As a means of determining wealth, work itself became largely irrelevant. Workers were alienated from their own labor and forced to sustain themselves through the marketplace by buying and selling representations of other people's labors. In this brave new world of colonial capitalism, which bears much resemblance to our own, social reproduction was dictated by the power to *represent* labor as valuable, not by the actual power of labor to create value.[15]

This book analyzes the interwoven histories of predation and production as colonial representations of work by examining the environmental practices and cultural contexts of raising and killing cattle, wolves, and other animals in the Northern Rockies frontcountry. It opens with the industrial fur trade of the mid-nineteenth century, exploring how wolfers, whiskey traders, and Blackfoot bison hunters created a mutual economic world based on the trapping and killing of wild animals. By the 1860s, federal officials in the United States and Canada had begun to circumscribe this trade as part of the larger task of preparing the frontcountry for agricultural production. They enforced laws prohibiting the sale of whiskey to indigenous communities and sought to confine those communities to reservations where Blackfoot bands would give up hunting for agriculture. In the meantime, the large-scale arrival of open-range livestock in the early 1880s altered regional relationships with wolves. The carnivores previously represented a means of production in the industrial fur trade, but after the arrival of livestock they existed as an anathema to the production of range cattle. From the 1880s onward, public and private authorities instituted a series of bounty programs to encourage the extermination of wolves and other predatory animals. Although largely ineffective for addressing the economic costs of livestock lost to wolves, this bounty system helped culturally and physically remake the frontcountry into cattle country by the end of the nineteenth century. The cattle industry emerged on the Blackfeet Reservation of northern Montana as well, bolstered by official federal efforts to transform the Blackfeet from hunters to herders, as well as by the routine trespass of local, non-Native cattle herds on reservation grasslands. Indicting Blackfeet hunting as a predatory leisure rather than a productive labor, the U.S. Office of Indian Affairs (OIA) built two reservation slaughterhouses as part of its efforts to stop the Blackfeet from "hunting" reservation cattle and to further consolidate agency control over what constituted appropriate work—in this case, wage labor for rations tickets rather than subsistence meat production. The early twentieth century witnessed a resurgence of Blackfeet resistance to these assimilation efforts. The community negotiated the individualizing imperatives of allotment by reinterpreting their social relations with the nonhuman world through a rejection of cattle ranching and a turn toward cooperative subsistence farming. Finally, this book closes with an exploration of how wildlife conservationists in the early twentieth century appropriated the logic of predation and production themselves to carve a place in western politics, framing their conservation projects as ranches for wildlife production and supporting the continuation of predator eradication programs into the 1930s.

The basic contours of the environmental and social inequalities of the North American West were structured during the period that this book analyzes. From the 1860s to the early 1930s, the West's extractive industries organized into large, corporate forces that dominated the region's land, labor, and water through most of the rest of the twentieth century. During this era, mining, timber, livestock, farming, and other enterprises fundamentally remade vast swaths of the western environment.[16] On both public and private lands, they reengineered ecological systems to better suit the accumulation of private profits with little regard for the short- and long-term consequences for local communities. With the assistance of federal, state, and provincial authorities, these enterprises restructured the landscape into a matrix of industrial uses in order to maximize production. In 1934 the historian Bernard DeVoto famously called the western states "America's plundered province."[17] Although the political power of these industrial forces has generally waned in our postindustrial "New West," its environmental legacies remain.[18]

The current controversies surrounding the return of gray wolves to the Northern Rockies are best understood as an ongoing consequence of these environmental and cultural histories of privatization and enclosure.[19] Today's coalition of ranchers and conservatives who commonly oppose the recovery and reintroduction of gray wolves in the region do so on the grounds that wolves threaten access to their private property. Wolves, they argue, can potentially steal the meat and money embodied in privately owned livestock either by killing those animals as prey or, as is more often the case, by stunting or delaying the physical growth of livestock by forcing them to graze more vigilantly.[20] From a legal standpoint, elk and other game animals live as public property, but the state practice of selling individual hunting permits effectively privatizes their bodies at death. Throughout much of the twentieth century and into the twenty-first, this relationship encouraged strategies of wildlife management similar to those employed by livestock growers. Federal, state, and provincial agencies raised and fed elk, bison, and other ungulates in a state of semidomestication; they created a system of so-called refuges to confine wildlife on slivers of public land within the region's broader fabric of private ownership; and they trapped, killed, and otherwise "controlled" carnivores in order to maximize their production of game animals.[21] The National Bison Range in Montana, the National Elk Refuge in Wyoming, and the defunct Wainwright Buffalo National Park in Alberta are three classic examples of this ethos of production at work in North American game management.[22] Studying the history of predator control in this context demonstrates that the extension of private property into western North America in the late nineteenth and early twentieth centuries not only carved boundary lines across the region's physical spaces but also institutionalized new property claims on the region's animals.

Owning an animal is never an easy or straightforward thing to do, however, since animals are acting subjects themselves and not simply objects to be dominated. Their own agency in our social encounters is revealed in our direct interactions with them and perhaps even more obviously in their decisions to avoid us. Owning an animal requires a certain denial of that capacity, an enclosure or confinement that maintains our access to their flesh or company while compromising their freedom. Even the love and affection of pet owners is necessarily posed within this structure of dominance.[23] The boundaries of parks and refuges further constrain the lives of animals we call "wild." Throughout the ongoing colonial history of our ownership society, we have had few solutions for those animals that actually leave the places we create for them—they are either killed or enslaved.

The historical practices that accompanied the creation of this world were experiences that often left little in the way of a paper trail. As a result, many of the arguments in this book are necessarily built upon wide-ranging fragments of historical evidence, and my historical interpretation is rooted in the analysis of those individual stories within their broader environmental and social contexts. Nearly ten years of extensive archival research went into the production of this monograph, and its historical evidence is drawn from manuscript collections spanning the Northern Rockies from Cody to Edmonton, as well as from documents housed at the U.S. National Archives, the Denver Public Library, the New York Zoological Society, and the Newberry Library in Chicago. The literal words "predator" and "producer" do not always appear in the historical record. Rather than focusing on these words alone, however, this book offers a history of predation and production as colonial representations of work by analyzing past environmental practices. Above all, it seeks to show how predation and production are more than just biological categories; they are also historical concepts rooted in the structures and contingencies of western conquest.

By investigating the historical role of predation in both justifying and contesting conquest and colonialism, this book also addresses broader ethical and philosophical questions at the heart of controversies over the exploitation of human and nonhuman life. At times, this book might seem unsympathetic toward livestock growers, but its purpose is not to wag fingers at those who earn their living by raising animals for slaughter. To the contrary, I hope that tracing predation's environmental and cultural histories will lead all of us toward more honest evaluations of our human footprints on the world. Maybe it can help us past the foolish question of whether we should or should not "prey" on living beings, as if we had the choice, and instead encourage us to consider how we can most respectfully and equitably sustain ourselves on the lives of others.

1

Wolves and Whiskey

If nobody got drunk, the East Coast would be awful crowded by this time.
—Charles M. Russell, "Whiskey"

Sprawling over twenty million acres of Montana grasslands lies "Russell Country," named by the state tourism board after Charlie Russell, Montana's most famous western artist. Bounded to the north by the Alberta border, the Rocky Mountains on the west, and dissolving toward the south and east across the grasslands of the upper Missouri River watershed, this sector of Montana's tourist geography spans an area three times larger than New Jersey. Buttes, cutbanks, and badlands lie beneath a big sky that dominates the horizon in every direction. The Marias, Milk, Sun, and Teton Rivers spill eastward from high mountain passes, emptying to the Missouri River as it muscles a muddy course to the city of St. Louis, two thousand miles distant.

The tourism board marketers have tried to domesticate this place by hawking its wildness as a consumable product. "If freedom could be bottled and sold," they declare, "the heart of Russell Country would be a popular place to acquire it."[1] But the earth's soul is not so easily peddled. It might be tempting to re-create the view from the glossy advertising inserts, to gaze across the frontcountry at sunset and imagine yourself as a pioneer conquering a virgin land. But the evening shadows of the Northern Rockies outrun your vision, and the breeze strikes a chill up your neck, engaging senses beyond the imperial gaze perfected by nineteenth-century landscape artists. In person, this place has a "physical ambiency," as Bob Marshall once described, an immediate and fluctuating beauty that exceeds the instantaneity of artistic representation.[2] Wallace Stegner described the sensation as a multiplicity of shifting "circles, radii, [and] perspective exercises," an overpowering display of light and movement driven by a wind "with the smell of distance in it."[3] Dan Flores called it "the visual equivalent of placing an ear to a seashell."[4] The sublimity of the high plains is powerful but subtle and unexpected. When you look at it, it exerts its own forces back; it crawls inside you. Russell Country is a place defined at these sensory margins, where grand visions surrender to the tactile and prosaic.

Russell earned fame as a painter but harbored a distrust of the medium. He understood himself as a storyteller first, and his brilliance hinged on discarding the static tropes of American landscape art for uncertain illustrations of rapidity and motion. Russell's canvases quiver with energy in ways that are absent from the work of his contemporaries, many of them easterners such as Frederic Remington and Charles Schreyvogel. Impatient wolves snort plumes of hot air as they trample circles through the snow. Horses kick with uneven gaits, and the earth shudders. In *Wolves at the Wagon-Train*, smoky campfires comfort a camp of plains freighters (figure 1). Three wolves watch from a distance, tails cocked in agitation, looking back at a landscape seen in reverse.[5]

If freedom came in a bottle, Russell would know. In 1927 he fused his explorations of place, paradox, and sensory confusion into a dissolute theory of conquest: America moved west because it was drunk. During the midst of the Prohibition Era, his essay on whiskey elevated the drink as an alternative heart of America's manifest destiny. With lighthearted humor, Russell wrote of drunk men passing out in St. Louis and waking up in Montana, "dragging a boat loaded with trade goods for the Injun country." No stranger to the pains and pleasures of whiskey himself, Russell celebrated the drink as a transcendent spirit of the western experience, a "brave-maker" that withdrew men and women from their inherent cowardice. He poked fun at the eastern moralists and reformers who tried to outlaw the liquor trade in indigenous communities in the nineteenth century and who passed the Volstead Act during the twentieth, a law that "made Injuns out of all of us." Like other mind-altering substances, rather than just "dulling the senses," alcohol provided humans with the possibility of expanding their comprehension of the world around them. The liquor laws were a disavowal of more significant sources of western misery that Russell witnessed during his lifetime. It was an age of unjust conquest that destroyed the wild freedoms of the northwestern plains and created Russell Country in its place, a matrix of private property and social exploitation created to enrich its new barons of land, grass, and cattle. It was funny to imagine drunken fortitude leading Americans westward, but its true roles in the region's ongoing colonial relationships were superficial. But beyond the humor, Russell had a serious point to make: "Whiskey has been blamed for lots it didn't do." Sobriety was just another false piety of colonialism and capitalism.[6]

Russell's observation rings true throughout the historical memory of the Northern Rocky Mountain frontcountry as well. The region's dispossession has long been linked with the whiskey trade, a transborder commerce that transformed the Blackfoot into "ragged mendicants," as the historian Hugh Dempsey described them.[7] The whiskey trade certainly connected many historical developments to one another. The arrival of the North-West Mounted Police, the Blackfoot's acquiescence to Treaty Seven, the near eradication of bison, and even innovations in steamboat technology all relied, in some respects, on the borderland's burgeoning nineteenth-century trade in whiskey, as numerous scholars have explored. Located at the center of these and other historical transformations, the spirit has taken on a life of its own. For many historians, whiskey has become an agent of conquest more potent than official actors like Colonel Baker or Colonel McLeod. In doing so, the spirit has provided a solvent to remove human contingency from the borderland's colonial past. Although the colonization of the North American West has most often been understood as a narrative of disenchantment, where Anglo-American concepts of science and modernity replaced indigenous spiritual practices, whiskey's emergence as the region's historical deus ex machina reveals that an unacknowledged animism also pervaded the borderland's western conquest. Viewed as a historical actor itself, whiskey contravenes a long-standing humanistic metaphysics that has reserved historical agency for human actors.

For many historians, then, it seems that the Blackfoot would not have been confined to reservations nor had their land and labor subordinated by the colonial nations of Canada and the United States were it not for the pernicious influence of firewater. Although alcohol has been controversial in many other encounters between Native and non-Native people, the exotic toxin seems to have soaked Blackfoot society more deeply, drenching it with a scale of death and destitution almost on par with epidemic diseases like smallpox. Between 1870 and 1875 hundreds, perhaps even thousands, of Blackfoot died through their contacts with whiskey, as Dempsey, Margaret Kennedy, Paul Sharp, and other historians have emphasized.[8] This wasn't the point that Russell had tried to make, but perhaps if nobody got drunk, the East Coast *would* be more crowded, or at least the conquest of Blackfoot lands might have stalled out farther east.

This chapter sets the groundwork for those that follow by arguing that the uneven acceptance of wolves and whiskey as historical agents demonstrates a defining but overlooked feature of colonial conquest: the need for colonizers to disavow their predatory inclinations. For federal officials and for the historians who later followed them, whiskey provided an opportunity for this disavowal by dehumanizing the borderland's narrative of conquest, shifting blame for Blackfoot dispossession to the mysterious alchemy of a spirituous liquor. But the historical outcomes of the whiskey commerce were far more complicated than often recognized. Intoxication did not just dull the senses; it also provided alternative and perhaps subversive ways of seeing, an acknowledgment long understood by the Blackfoot and other indigenous people of the Great Plains. Over generations, the Blackfoot had developed a sophisticated palette of skills for interacting and communicating with their nonhuman kin throughout their dreaming and waking lives. In addition to ritual fasts and various ceremonial procedures, the consumption of liquor and other substances helped open these lines of communication between human and animal worlds. Colonialism comprised the process of melting away these kinship ties in order to subordinate Native land and labor and to establish the frontcountry as an ordered domain for industrial agriculture. The U.S. Army and the Office of Indian Affairs were tasked with prohibiting Native access to liquor as part of this process.

1. Charles Marion Russell, *Wolves at the Wagon-Train*. Courtesy of Paul E. Sharp, *Whoop-Up Country: The Canadian American West, 1865-1885*, 2nd ed. (Norman: University of Oklahoma Press, 1973).

The chapter is subdivided into two sections. The first explores Blackfoot understandings of nonhuman kinship and assesses the role of the whiskey trade in early colonial efforts to control the "Whoop-Up Country" of present-day northern Montana and southern Alberta during the 1860s and 1870s. The second section explores the effects of the whiskey trade on wolves and on wolves' roles in establishing the frontcountry as a predaceous landscape both culturally and physically. The whiskey trade did not play as significant a role in destroying Blackfoot life and livelihood as previous historians have concluded. Rather than crippling the Blackfoot, the trade emboldened Native resistance to colonization, which is one reason why American and Canadian authorities sought its dismantlement. Another important outcome of the whiskey commerce was a surplus of bison carcasses scattered across the frontcountry, skinned for the robes by Blackfoot market hunting. Wolves fed on these carcasses and most likely increased their populations through the end of the 1870s. Their increasing presence alone indicated the frontcountry's transformation into one of the colonial periphery's most predaceous landscapes.

Whoop-Up Country

In the fall of 1869 two white men traveled north from Montana across the forty-ninth parallel with a wagonload of whiskey and lever-action rifles. They did not know where exactly they had crossed the international boundary into Canada, but they did know that the place they built their trading post, at the confluence of the Oldman and St. Marys Rivers, was far enough from Montana to avoid harassment by the U.S. Army. Over the winter, they struck a brisk trade with Blackfoot hunters who eagerly exchanged hundreds of bison robes for the new repeating guns. Whiskey proved a popular commodity too, as it had across the border, until earlier that summer, when U.S. marshals and Office of Indian Affairs agents had broken up the illegal commerce. The next spring, John J. Healy and Alfred B. Hamilton returned to Montana and sold their robes at Fort Benton, splitting a $5,000 profit. For the next few years they ventured to the same location above the border, building a permanent log structure after their first burned to the ground and increasing the scale of their trading operations, bankrolled after 1870 by the town of Fort Benton's leading merchants. Unmolested by federal authorities, they traded whiskey freely. When asked for an update on business, Healy responded, "We're just whoopin' up on 'em!" an exclamation quickly recruited as the name for the trading post, Fort Whoop-Up, as well as a nickname for the broad sweep of Blackfoot country from the Missouri to the Saskatchewan Rivers.

The year 1870 marked the start of a commercial revolution on the high plains straddling the forty-ninth parallel. From Fort Benton, the highest navigable point on the Missouri River, industrially produced goods began flowing northward across the international boundary, offering the three tribes of the Blackfoot Confederacy—the Pikuni, Kainah, and Siksika—unprecedented access to manufactured clothing, repeating rifles, canned food, and a host of other items, including whiskey, most notorious of them all. Blackfoot trade with whites, especially with white Americans, picked up substantially over the course of the next decade, and by the early 1880s the commerce had generated a handful of fortunes across northern Montana. The merchant houses of T. C. Power and I. G. Baker benefited the most from the trade and consolidated their market power with dramatic expansions eastward, breaking apart the old fur monopolies of Hudson's Bay, St. Louis, and St. Paul in the process. With satellite offices in Chicago, Montreal, and New York and with vertically integrated operations to streamline their flows of profits back to Fort Benton, the merchant houses helped momentarily reverse Montana's usual arrangement under core-periphery relationships. Rather than a dusty frontier outpost, Fort Benton truly was "the world's innermost port"—the center of a global traffic in bison robes and wolf pelts.[9]

The furs of these animals had superseded beaver skins in profitability by the late 1860s, a transformation based on a series of shifting political, ecological, and economic circumstances, along with changes in Victorian fashion sensibilities. In many parts of the Northern Rockies, especially drainages trapped heavily for export by the Hudson's Bay and American Fur Companies earlier in the century (the mountain streams west of Edmonton and those southwest of Helena, respectively), beaver populations had plummeted to a historic nadir.

The Blackfoot were bison hunters first and warriors second, a reality that had frustrated Hudson's Bay Company (HBC) traders for nearly two centuries, ever since their first encounter with the powerful tribes in the 1690s. Even by the 1860s the Blackfoot remained aloof to the HBC's requests for small-game furs, instead bringing hundreds of bison robes to HBC trading posts during years when they decided to come at all. Still reliant on canoe and portage routes through the Canadian Shield, the HBC could not efficiently transport these massive bales of furs. Moreover, the Blackfoot continually warred with the Crees and the Assiniboines, the HBC's prime suppliers of beaver.

Meanwhile, the 1860s also witnessed the near collapse of St. Louis's American Fur Company (AFC), another heir to the beaver's early nineteenth-century slaughter. Like the HBC, the AFC was forced to accept the Blackfoot's insistence on trading bison robes, a commodity the AFC, initially, did not particularly want. But armed with steamboat navigation of the upper Missouri River, by the late 1830s the AFC had come to the trading table, eager to ship Blackfoot bison robes downriver to its St. Louis entrepôt. From its post at Fort Union, at the junction of the Missouri and Yellowstone Rivers, just east of the present-day Montana–North Dakota border, the AFC dominated the fur trade of the northern plains. But although profitable, the Blackfoot bison robe trade was not great enough to offset the AFC's losses on other furs. Moreover, the firm lacked the capital necessary to expand its business with the Blackfoot by supplying more trade goods. By 1865 a number of competing firms working out of the town that had sprung up around the AFC's Blackfoot trading post, Fort Benton, had eroded AFC profits to the point where the company decided to sell its post on the upper Missouri. It found a willing buyer in the Northwest Fur Company, headquartered in St. Paul, Minnesota, which similarly suffered from competition, selling out to I. G. Baker by the end of the decade.[10]

Throughout the middle of the nineteenth century, the Blackfoot drove a hard bargain from these firms, demanding sophisticated items like rifles, finished clothing, saddles, and metal ware. Because it was ruled illegal in both the United States and the Dominion of Canada, liquor played a minor role in the Blackfoot trade until the late 1860s, when independent Fort Benton traders, operating with less judicial oversight than the large firms, began trading whiskey from small, decentralized posts scattered across northern Montana. Contrary to popular understanding, however, the Blackfoot were already well acquainted with liquor by this period. The HBC had periodically dispatched rations of rum as ceremonial trading gifts to the Blackfoot since the eighteenth century. In the 1830s the AFC had actually smuggled a large still northwest from St. Louis, transporting and reassembling it at Fort Union. The still churned out trade whiskey for several years before its discovery by a U.S. Army officer on his travels across the plains. Faced with the possible revocation of its trading license, the AFC destroyed the massive still while claiming it was reserved solely for the use of AFC personnel. Thomas Hart Benton, a U.S. senator from Missouri and an AFC board member, facilitated a settlement whereby the AFC merely paid a fine for its transgression and continued its business. The AFC named its trading post with the Blackfoot, Fort Benton, in honor of this senator who saved the firm from the full wrath of U.S. commerce laws.[11]

By 1870 the Blackfoot had become well versed with whiskey and its intoxicating effects, having traded for it in relatively lesser quantities for at least two generations. Also by this time, whiskey was also neither incompatible with nor unprecedented within Blackfoot cultural practices. Liquor had long played a significant role in the gift giving that often accompanied trade with both the HBC and the AFC. Furthermore, although it is unlikely the Blackfoot had access to alcoholic drink before their encounters with white traders, they did have significant experience with other intoxicating substances, especially body-purging teas. Sometimes the Blackfoot consumed these teas in large quantities during ceremonies in which participants would try to outdrink one another before getting ill.[12] The incorporation of whiskey into these ceremonies was more a change in means than in ends.

The anthropologist Clark Wissler, who studied Pikuni environmental and cultural practices and conducted fieldwork on the Blackfeet Reservation of northern Montana during the early twentieth century, theorized that the Blackfeet tea dance was probably acquired from a similar but more formal Pawnee ritual from the southern plains.[13] As Wissler's informants described the ritual, competing teams of Blackfeet drinkers lined up opposite one another and then presented their opponents with large containers of tea, which they had to consume. The contest ended when one side finished drinking its tea or when the other side had to forfeit from feelings of "sickness," ranging from bloating to vomiting. According to a Blackfeet man named Curly Bear, whom Wissler's Native assistant David Duvall interviewed, the tea dance was remarkably informal, with no leader "nor any ceremony of any kind."[14] This open drinking format would have been mutually comprehensible to both Blackfeet and non-Native newcomers steeped in the tavern and drinking cultures of the nineteenth century.[15]

The historical origins of Blackfoot tea-drinking rituals are unclear, but evidence suggests they may be related to broader Native practices from the precolonial Great Plains that were focused around the ingestion of psychotropic plants such as peyote, the small cactus that grows only in portions of the Chihuahuan deserts of northern Mexico and southern Texas. It is unlikely that the Blackfoot, like other indigenous people of the Great Plains, often, if ever, had direct access to fresh peyote in the course of their travels, but they almost certainly traded for it with neighboring Native communities and had historic experiences with the plant's psychoactive effects.

Although commonly eaten, peyote can also be consumed when dried and preserved by soaking it in water and drinking the liquid.

Wissler documented Blackfeet men singing transcultural peyote songs in 1913 but also determined that "the peyote cult seemed not to have reached them."[16] For its part, the Office of Indian Affairs officially denied the existence of peyotism among the Blackfeet, and the Native American Church likewise listed the Blackfeet as "non-users" in 1922.[17] But the dried flesh of the plant made its way through Native trade networks far from its desert origins, likely reaching the Lakotas, Cheyennes, and even the Blackfeet's Cree neighbors by the early nineteenth century. Perhaps the dogmatic religious aspects of southern plains peyote consumption did not appeal to the Blackfeet, but in all likelihood they were knowledgeable about peyote and had used it themselves for nonreligious purposes.

In any event, like most other Native communities of the Great Plains, the Blackfoot had long used various techniques to "intoxicate" themselves as a means of transcending their human selfhood and interacting and communicating with an other-than-human world. At first blush, it might seem unsophisticated and embarrassing to consider the possibility that humans harbor a suppressed animal nature lurking trapped behind psychological walls erected in historical time by religion, capitalism, and other forms of social discipline. But to explore that possibility offers some insights into western history, namely, that the colonial process was part of this broader effort to consolidate power by forcefully limiting reality to conform to a human economy dissociated from its nonhuman context.

Within Blackfoot ontology, basic categories of human and animal held a significance very different from the race- and species-based taxonomy maintained by colonial society. Like many other indigenous cultures in North America, the Blackfoot historically understood their collective identity on the basis of kinship rather than race. But kinship in its usual sense of heredity also insufficiently grasps the exact nature of Blackfoot social ties, which included kinship relations between individual human and nonhuman persons. Constituted more by impermanent alliances than by immutable lines of sexual descent, kinship within the so-called Blackfoot Confederacy never sparked the creation of a discrete political body but was instead a way of comprehending one's relations with human and nonhuman others. The Niitsitapi—or the "real people," as the Blackfoot called themselves as a collective—included various nonhuman animals along with human ones. To be "real" was, in these terms, a matter of relationship rather than of purity. The real people were real precisely because of their interspecies kinship, not because they thought of themselves as the only real people in a world filled with inferior, animal-like others.

In Blackfoot country, then, Blackfoot people generally agreed that they were animals—or, at least, that they were not *not* animals. Abilities to hold social relations with nonhuman animals were understood as sources of power rather than evidences of stalled human evolution. Take, for example, the interesting case of the Wolf Man as related by a Blackfoot storyteller to George Bird Grinnell in the 1890s: A man finds himself trapped in a pit but is rescued by a wolf, coyote, badger, and fox. Realigned with these creatures, with whom he can suddenly communicate, the Wolf-man goes on to help these carnivores steal meat from traps, although he retains his human form. For a time, the Wolf-man lives like a wolf, but he is still a man. When he is caught stealing meat, his former human kin recognize him, and so the Wolf-man rejoins his former human alliance.[18]

The temporary mutability of a human marker like language in this story and its exchange for other powers confounds a colonial taxonomy that abstracts communication as a property of a particular species rather than a practice of interspecies engagement. In this sense, human-animal difference, according to Blackfoot understandings, does not exist as an abstract separation but is established through lived relationships between human and nonhuman persons. Difference is forged through the process of trying to become animal, to adopt nonhuman powers, to exist within alliances of humans and animals. It is a process best understood, within Western philosophy, perhaps, as a relationship of mimesis and alterity—the maintenance of difference through mimicry—as Michael Taussig, Rane Willerslev, and other anthropologists have explored.[19]

Blackfoot scholars such as Betty Bastien have identified colonization as an attempt to replace this ontology of natural alliance, as she calls it, rooted in practice, experience, and geography, with a Western taxonomic ontology that reduces kinship to heredity. She writes that, under these circumstances, colonization is accomplished "by redefining identity, self, and humanness as abstractions, instead of defining them through the specific lived realities of natural alliances."[20] The resulting destruction works out in a twofold process: first, by dissociating Blackfoot connections to the nonhuman environment, and second, by instituting a Western model of autonomous selfhood that undermines Blackfoot obligations to community.

The first colonial challenge to Blackfoot natural alliances played out in the near physical destruction of the northwestern plains' bison herds, a process complete by the end of the 1870s. With the destruction of the bison, the Blackfoot lost not only a means of subsistence and a means of trade but also, more significantly, a way of establishing relations with nonhuman others. Without identity-forming relations with bison and other animals, notions of Blackfoot kinship were cut adrift. Furthermore, with the eventual establishment of reservation geographies and allotment and enrollment procedures in the United States and Canada, Western genealogical understandings of kinship strained natural alliances as a politically expedient form of collective Blackfoot identity. This move to a genealogical model also cemented an abstract dissociation of Blackfoot people from Blackfoot land.[21]

The second set of colonial challenges was the institution of autonomous selves at the core of Blackfoot identity. As Bastien postulates it, the Canadian government (along with the U.S. government) sought to accomplish this transformation through assimilation policies like boarding schools and rations systems (and, of course, allotment in the U.S. case), which worked to replace tribal obligations with a sense of capitalistic individualism. This replacement, according to Bastien, "has created unprecedented conditions of dependency . . . the Third World conditions in which many Aboriginal people find themselves in Canada."[22] As we shall see in chapters 3 and 4, the histories of the cattle industry and land allotment on Montana's Blackfeet Reservation demonstrate that it was difficult, if not impossible, for Blackfeet individuals to succeed as autonomous selves within a colonial system that poured Native people into the capitalist economy while restricting their basic rights, such as mobility and the ownership of private property.

Whiskey most likely provided a vehicle for reconstituting these natural alliances in the midst of their colonial dismemberment. Just as precolonial ceremonial practices offered mood-altering settings to transform Blackfoot perceptions of the world around them and to expose their connections with their nonhuman kin, whiskey accomplished a similar chemistry. Liquor may have been a new addition to the Native repertoire in this respect, but its uses as a spirit fell within established traditions. When colonial administrators restricted Blackfoot access to liquor, it was much like later efforts to curtail Native rituals such as the Sun Dance; it was part of a larger prerogative to constrain indigenous practices that sought to transcend the sober divisions and dissociations of the colonial state. Similar to peyote and other psychotransformative substances, liquor posed a threat to the unraveling of indigenous nonhuman kinship alliances.

Blackfoot communities, of course, were not the only ones drinking liquor. Whiskey traders sold their illicit goods to the very soldiers tasked with policing the trade. Beset with boredom, cold winters, and easy access to alcohol, troops in the Whoop-Up Country drank themselves into a stupor. Charged with maintaining the peace by prosecuting Indian liquor violations, the soldiers themselves, including their officers, drank heavily. Medical records from Fort Shaw indicate that in the winter of 1869–70 soldiers reported "unfit for duty" averaged from 6 to 12 percent each week. Since other categories existed for illness and injury, it is reasonable to assume that a substantial number of troops were consistently drunk or hungover. Several explicit entries in the casualty listings reference liquor as a factor in death. In one instance, "Private Patrick Stanton, 13th Infantry," the surgeon recorded, "was found dead March 4th, 1870 some three miles above the Post. He had wandered off the evening before while intoxicated and perished from exposure."[23]

Evidence from military and business records indicates that whiskey traders struck a brisk business with the troops. In 1871 the commanding officer at Fort Shaw demanded that nearby trader Joseph McKnight stop selling liquor to the soldiers. The text of the reprimand dictated that McKnight stop selling "larger quantities than by the drink, to be taken at the bar, except upon written permission of the Post Commander."[24] A pragmatist, McKnight switched to beer in order to satisfy this directive and still satiate the soldiers. In the interim, however, he continued to sell bottles of whiskey in private transactions to both soldiers and officers who forged special dispensation permits. McKnight kept these slips of paper, probably to cover his tracks in case of another reprimand, and a number of them survive in his business books.

Evidence like these slips might be important to keep track of, since the year before McKnight's arrival, the fort's commanding officer arrested Walter Cooper, a trader whom the colonel caught trucking kegs of whiskey right outside the fort. Colonel Reeve confiscated Cooper's horses, wagon, and whiskey and imprisoned him overnight in Fort Shaw's jail. Fortunately for Cooper, Fort Benton whiskey interests ran the law in northern Montana. As one officer lamented, "There is a community of interest among the capitalists here that enables them to combine and have most things their own way in many respects."[25] During Cooper's trial, the county prosecutor failed to show up on Colonel Reeve's behalf, and the jury refused to indict Cooper on any charges. Cooper's attorneys immediately fired back with a personal lawsuit against Reeve and also a suit against the army for wrongful imprisonment.[26]

Local suspicion and hostility toward federal authorities flared in many other ways as well. The majority of Bentonites had emigrated from the southern and border states during the Civil War. Many were Democrats, some were deserted soldiers, and most opposed the authoritarian and militaristic policies established by eastern governments in the 1860s and 1870s to control western settlement and Indian affairs.[27] The Conrad brothers, I. G. Baker's lieutenants in the liquor trade, were former members of "Mosby's Raiders," a paramilitary guerrilla force that fought General Philip Sheridan's cavalry in northern Virginia during the closing days of the Civil War. It was not surprising, then, that Fort Benton's participants shunned federal and territorial authority and traded whiskey freely. Hostile antiauthoritarianism in the Whoop-Up Country played out in interesting ways. As early as 1867, tensions between the Blackfoot and Sun River ranchers had convinced Acting Governor George Meagher to organize a militia. Coming from the mining and ranch land of southern Montana Territory, Meagher probably did not expect the hostile greeting he received at Fort Benton. In July, shortly after arriving to pick up a shipment of rifles and ammunition sent upriver by steamboat, he mysteriously drowned.[28]

Earlier that year, tension mounted between Fort Benton's merchants and the federal government after soldiers seized one of I. G. Baker's wagon trains moving whiskey across the Blackfeet Reservation. The U.S. secretary of the interior chided Baker, informing him that "the laws . . . are undoubtedly in force within the territory in which the Indian title has not been extinguished." Baker boldly responded: "If Hon. Secy. Of the Interior believed [this], he has been derelict in duty, in not driving the thirty thousand settlers of Montana from the homes they are unlawfully holding on Indian lands."[29] The success of Baker's trading operations rested on the Blackfoot's ability to hunt bison over the huge expanses of the northwestern plains. Baker opposed federal power in the Whoop-Up Country not only because it prosecuted his illegal trade but also because it tolerated illegal homesteaders and other agricultural migrants who threatened to displace the Blackfoot workers who were vital to Baker's enterprise.

In this manner, antagonism between Fort Benton merchants and the federal government worked to postpone settlement. While the Blackfoot fought off white intrusion on their homeland, I. G. Baker and T. C. Power battled expansionist federal agents in local and national courts. Local white businesses in the Whoop-Up Country were not always complicit with federal attempts at Indian removal. Until the arrival of cattle into the region in the following decade, the federal government represented the primary colonial force in the Whoop-Up Country of the 1870s.[30]

But this trade in whiskey and bison robes did leave a harmful mark on the Blackfoot. The unbalanced terms of this trade reoriented Blackfoot labor toward market production, accelerating their annual kill of bison, which destroyed traditional Blackfoot subsistence and impoverished the tribe.[31] Moreover, this economy of whiskey and bison robes operated in similar ways to the late nineteenth-century meat industry.[32] Like the meatpacking elite of Chicago, Fort Benton's merchant class harnessed the natural wealth of its northern hinterland for industrial demand through layers of intermediary labor and technology that hid environmental and social costs from alienated consumers. Unsurprisingly, contemporary Bentonites referred to their city as the "Chicago of the Plains."[33] The transformation of bison into robes occurred largely with Native spaces, but the transformation of robes into money took place within the stone walls of the Fort Benton merchant houses.

Whiskey played an important role in the trade because it helped reorient the Blackfoot social structures that dictated restraint over hunting practices. The Blackfoot's remarkable recalcitrance to trade had impressed and angered white traders throughout the early nineteenth century. By the 1850s, however, the Blackfoot had slowly integrated into a white market economy, trading robes with Fort Benton's AFC and the British HBC. During this early period of trade, the Blackfoot considered bison hunting to be mainly a communal effort. The hides and tongues of dead bison belonged to the hunter if he desired to keep them, but the meat and other remains were distributed amongst the tribe.

To female labor fell the unenviable task of skinning and butchering. While the Blackfoot traded in communal groups with the HBC and AFC representatives at their centralized posts, the decentralized whiskey traders from Fort Benton who had superseded them by the early 1870s catered more to smaller groups of Blackfoot hunters and their immediate families. Thus, Blackfoot hunting gradually changed in orientation from an activity of larger bands to one of smaller family units. Some historians claim that this transformation was evident in changing marriage patterns. Men started marrying a greater number of wives, and by the late nineteenth century, the average age of brides had dropped from around eighteen to as low as ten years old.[34]

It is also worth considering how transformations in bison disease ecology during the 1870s intensified pandemics among the Blackfoot. By the 1870s horses and livestock had likely transmitted brucellosis to the northwestern bison herds.[35] Undiagnosed until the early twentieth century, this Old World bacterial disease can cause miscarriage in pregnant women who have not developed immunity. It is possible that Blackfoot women contracted the disease, which is transmitted to humans through the inhalation of bacteria from infected animal tissue, during their daily gutting, skinning, and fleshing of bison robes.³⁶ Although clearly overshadowed by the devastation of smallpox, brucellosis was likely a daily source of misery on the plains that deserves more research from future historians.

By the early 1870s the Blackfoot had beaten back the Crees and the Assiniboines in the north, opening further hunting ground toward the east. Part of their success was due to their close connection with Fort Benton traders, who illegally traded hundreds of state-of-the-art repeating rifles to their Blackfoot allies. In 1870 two bands of Pikuni and Kainah camped near Fort Whoop-Up routed six hundred Cree and Assiniboine warriors, killing nearly half of the larger force with a steady barrage of accurate rifle fire.³⁷ In this way, the whiskey-trade years marked a brief expansion of Blackfoot territory in the early 1870s.

The inebriating effects of whiskey created relatively little hardship and grief for the Blackfoot. Rather, the cumulative ecological and social effects of the whiskey trade itself wreaked havoc. Although drunken spats and alcohol poisoning undoubtedly killed some men and women, disease and the destruction of subsistence dwarfed these episodes in importance.

Wolves and the Ambiguities of Predation and Production

With the Whoop-Up Country's bison herds left slaughtered and skinned on the plains, wolves proliferated. Their abundance laid the foundation for "wolfing," a method of poisoning wolves by baiting carcasses with strychnine. After spreading these baits over a wide swath of prairie, wolfers rode out to collect their kill every few days, sometimes returning with dozens of dead wolves. Once back at their base, usually a crude shack on a river bottom built to accommodate two to four men, the wolfers performed the gory labor of gutting and skinning. Compared with the small whiskey traders, wolfers could earn surprisingly good money for their efforts after selling the pelts downriver in Fort Benton, although major risks also accompanied their endeavor. As in the robe trade, however, the main profitability of wolf pelts rested in their freight and distribution within an international market system, and T. C. Power and I. G. Baker were the primary beneficiaries. Unlike the whiskey enterprise, though, the profound devastation wreaked by wolfers did not lead to a local spirit of mutualism between these bands of white men and the Blackfoot. On the contrary, reciprocal bouts of violence erupted between these two factions, greatly complicating local politics and bringing the whiskey trade to an abrupt end.

In 1870 approximately 22,500 wolves and 450,000 bison thrived in the Whoop-Up Country.[38] However, the immense bison slaughter provided an unprecedented food source for wolves. With access to good nutrition, female wolves can breed a year earlier and mother twice as many pups.[39] Wolves' remarkable numbers and their adaptability impressed human observers throughout the 1870s. Learning to follow the sound of gunfire, wolves surrounded hunters and waited for their opportunity to scavenge fresh meat. In the late 1870s Dan McGowan reported that he had "never seen gray wolves so numerous as now. When we are skinning and cutting up the buffalo they form a circle around us and wait impatiently. As we move away," he continued, "they rush in to fight over the offal. Many wild fights are witnessed but ammunition is scarce and we refrain from shooting."[40] Under these circumstances, wolves became easy prey for trappers.

Not only did the slaughter of bison increase wolf reproduction rates, but through their annual kills, wolfers also reduced any overpopulation pressures that huge wolf populations might otherwise have faced. Recent studies throughout Canada and the United States have concluded that wolf reproduction rates reach easily into the range of 200 to 400 percent following extensive population die-offs and that wolf populations can sustain annual human-caused mortality rates anywhere from 35 to around 74 percent.[41] Thus, the combined ecological effects of increased bison mortality *and* wolf mortality caused wolf populations to surge to incredible levels. Even if wolfers managed to completely wipe out a local range of wolves, in-migration and increased reproductive rates would quickly reestablish the population. Estimating from T. C. Power & Co. shipping files, between 1871 and 1875 wolfers killed around 34,000 wolves in the Whoop-Up Country.[42] This level of mortality would have been easily sustained by wolf populations but also was probably high enough to stimulate increased reproduction. Wolfing, then, did not serve to control predator populations but instead increased wolf numbers.[43]

Stereotyped as dumb and uneducated by historians and generally reviled as poor and uneducated by the northern plains' merchants and ranchers, a surprising number of Montana's early bourgeois men actually got started by killing wolves as recent immigrants. In this respect, wolfers were not unlike the upwardly mobile "mountain men" described by the historian William Goetzmann.[44] Wolfers could and often did make modest sums of money that they could later parlay into other enterprises. In 1867 Charlie Rowe reported that some friends who wolfed near the Milk River "got hides worth $2,000 or $3,000 in few months and did not work hard or steady."[45] Oscar Brackett's telling reminiscence depicts the transitory and offhand flavor of wolfing. "I got tired of Cutting Wood," he complained,

> so I and Archey McMurdy [and] Bill Hook took our Poneys, packed them with supplies & started up Sunday Crick. We was going to make our fortune poisning Wolves. . . . We did not get many Wolves about a hundred big ones. The two Johnson & one Jackson with two others got fifteen hundred Wolves that winter, they took big Circles and put out lots of poison, Wolves at that time went in big Bands & followed the Buffaloo up. It was common thing to poison 40 or 50 at one bate. We got 34 All at one time. . . . I stade at the Cabbain until the river broke up. McMurday & Hook went up the River have never seen them since.[46]

If the other wolfing party came away with 1,500 pelts, they could probably sell their winter's labor for at least $3,000, or $600 per person. As for Brackett, he probably made less than $70 that winter. In comparison, George Clendinnen paid Peter Koch $75 a month to take charge of his trading post on the fork of the Missouri and the Musselshell Rivers in 1869. Most wolfers achieved mixed financial results. Koch wrote his uncle explaining that in winter, Musselshell was the "dullest of dull places, where nothing at all is going on. Everybody has gone out wolfing." He also observed that "there are very few men here that save anything. All the summer they do little or nothing besides drinking whiskey and gambling and when winter comes, they are generally in debt so deeply for whiskey and provisions, that it takes half the winter to get clear of that."[47]

In addition to whiskey, wolfers purchased strychnine in great quantities, sprinkling its deadly granules on their bait's tongue, sliced-apart rump, and spilled entrails. This chemical, isolated and concentrated into a mass-produced, granular form, wreaked absolute death on the plains and laid waste to life in a manner that outraged the Blackfoot. It was not simply the wanton destruction of wolves that enraged the Blackfoot but rather the wider swath of death and destruction that wolf baits spread across their hunting ground. Strychnine-laced baits killed wolves but also immense numbers of other plains scavengers, including coyotes, foxes, birds, and domesticated dogs. The death of countless dogs particularly irritated the Blackfoot. Furthermore, the spastic, toxic vomit these poor creatures heaped upon the earth during their final moments poisoned grass and killed horses and bison.[48]

The hostility that developed between wolfers and Blackfoot led to increasingly brutal reprisals. Blackfoot routinely attacked and killed small groups of wolfers who trespassed on their hunting ground. Occasionally, wolfers took the upper hand in these engagements, but more often the Blackfoot outnumbered and, armed with the latest repeating rifles, usually outgunned the wolfers. In 1872 Thomas Hardwick, a former Confederate soldier and prolific murderer in the Whoop-Up Country, led a group of sixteen wolfers near the Sweetgrass Hills. When approached by a band of peaceful Assiniboines, Hardwick and his followers suddenly opened fire, killing four of them. The next spring, while bringing pelts to Fort Benton, a Blackfoot raiding party surprised Hardwick's group and stole their horses. Chasing the thieves around the plains with no success, Hardwick's frustrated group attacked Assiniboines again, this time surprising a band camped outside a whiskey trading post in the Cypress Hills run by a well-known trader named Abel Farwell. Farwell and his Crow wife tried to prevent the slaughter but barely escaped murder themselves. The wolfers killed over thirty people and demolished the camp.[49]

Animosity also flared between the wolfers and the whiskey traders, the source of the Blackfoot's powerful weaponry. Prior to the Cypress Hills massacre, a wolfer named John Evans put together a band of wolfers to deliver an ultimatum to John Healy at Fort Whoop-Up, demanding that he stop trading guns and ammunition to the Blackfoot. Healy discovered the scheme in advance and greeted Evans's arrival with a loaded cannon.[50]

Conclusion

During the 1860s and 1870s the Montana-Alberta borderland's emergence as a landscape of predation was both a cultural and a material development of its history of conquest. For Anglo-American wolfers and whiskey traders, the region was indeed a "happy hunting ground" where they could indulge in enterprises deemed predatory by the colonial governments of both the United States and Canada. And the ecological effects of the whiskey, bison robe, and wolf pelt trades actually created an environment disproportionately populated by predators, especially wolves, whose opportunistic population dynamics thrived on the region's abundance of carrion. In the midst of these socio-ecological transformations, predation became the pariah of the American and Canadian nation-states, both of which sought to enact colonial visions of agricultural settlement and production on the northwestern plains. Purging the land of predators, both human and animal, became the mantra for federal officials north and south of the international boundary. The consequences of this project fell most harshly on the Blackfoot, whose hunting traditions were circumscribed by the U.S. Army and the North-West Mounted Police and who, by the end of the 1870s, were confined to the Blackfeet Reservation in Montana and the Blood, Piegan, and Blackfoot Reserves in Alberta. It also fell on the fledgling Fort Benton merchants, a population of white men derided as an anathema to civilized settlement who made their fortunes plying whiskey, guns, and other goods to the Blackfoot in exchange for products of the hunt.

Although whiskey has played a vital role in many histories told about the Montana-Alberta borderland, wolves have been common as metaphors and images of the western landscape's predaceous environs but less often considered as historical subjects of analysis in their own right. Look again, for example, at Russell's *Wolves at the Wagon-Train*, which Paul Sharp published as the frontispiece of *Whoop-Up Country* (figure 1).[51] It is unclear why the train is circled or what its wagons carry, maybe whiskey traders, maybe wolfers, maybe something more benign? Three wolves observe the vague scene from a distance, their tails somewhat agitated. If they are the harbingers of destruction, over whose flesh are they salivating? Who is the predator, and who is the prey? Wolves were much more than simply predators—and as cattle increasingly arrived to the frontcountry in the 1870s and 1880s, wolves would not only exist as obstacles to the production of beef but remain themselves a means of production.

2

Beasts of Bounty

During the forty-year rise and fall of the northwestern plains' open-range stock industry, "wolfing" was big business on the Montana-Alberta range. By the 1870s this regional term had come to designate the human labors of trapping, poisoning, or otherwise killing wolves throughout what is now northern Montana and southern Alberta. Originally hunted for their furs, which trimmed fashionable Victorian jackets, wolves later became bountied animals for their depredations on Montanan, Albertan, and Blackfoot cattle herds. Government agencies, agricultural associations, and large stock growers offered between two and a hundred dollars apiece for wolves' so-called scalps, incentives that led to the deaths of hundreds of thousands of the animals in Montana and Alberta alone from the 1880s to the 1920s. This labor of extermination fell mostly to the poor, the unemployed, or the colonized. Animalized themselves as predators, wolfers stalked and killed wolves to reengineer the range for the needs of colonialists and capitalists, labor that further pushed wolfers toward the margins of colonial society.

This chapter traces the transformation of wolfing during this period, roughly 1860 to 1920, from its origins in the late industrial fur trade to its professionalization and marginalization under the yoke of the livestock industry and the bounty system. Wolfing was not the straightforward result of Anglo-American prejudice against the animals from time immemorial, nor did wolves and stock raisers face inevitable conflicts with one another that justified the removal of one or the other. Instead, the practices and consequences of wolfing in the frontcountry unfolded within a specific colonial project of consolidating Anglo-American claims to the land on the basis of designating differences between productive and predatory forms of labor. The image of the open-range Anglo-American cattleman epitomized the first form: taming the northwestern plains, producing and reproducing value through the protective husbanding of domestic stock, mostly beef cattle. Of course, this image was an illusion. Until the widespread adoption of sustainable mixed farming and winter haying in the early twentieth century, the range cattle business on the northwestern plains destroyed more fortunes than it created. But the illusion stuck, and by the 1890s the range cowboy had become synonymous with production on the plains, a development that relied on the cocreation of predators.[1]

Wolves and wolfers—along with Blackfoot, Basques, and others discussed in subsequent chapters—embodied this predatory figuration. According to colonial representations and material experiences of labor and value, these alleged predators skulked along the range's wild frontiers, seizing sustenance from the frontcountry's so-called legitimate producers and improvers. Wolves proved particularly troublesome, because as nonhumans, they were not so easily subjected to domination through colonization's political economies of race and class. Although the colonial elite could mandate wolves' extermination, they could never effectively chasten wolves into the service of capital accumulation. Moreover, profiting from killing wolves under the bounty laws required a disavowal of conventional production. In contrast to the productive labors of ranching, killing wolves involved instead a metamorphic "becoming-wolf," an overt and violent participation in a predatory economy of stalking and killing, an occupation increasingly left to those on the colonial margins.

Cattlemen sought to disavow the predatory nature of their own "productive" labor in order to legitimate their ownership claims on the frontcountry of the Northern Rockies. They justified their own seizure and exploitation of life, land, and labor on the basis of representing wolves as illegitimate harvesters of animal flesh. At the same time, they understood their own efforts to raise and sell cattle for an industrial beef market as a form of productive labor. But the lines that separated productive and predatory labor in the minds of stock raisers were muddled and subverted by the environmental practices and social consequences of wolfing. Both cattlemen's and wolfers' personal interactions with wolves revealed that wolves were beasts of bounty in multiple senses, animals that were both obstacles and means to realizing value. Stock raisers loathed them as vicious, anthropomorphic murderers, while many wolfers recognized them as old friends, especially in the cases where wolfers raised wolf pups in captivity. These everyday intimacies constituted an animated world where the lines between human and nonhuman fluctuated but still retained their potency. Caught within cultural anxieties over production and predation, humanity and animality, wolfers drifted toward the margins of colonial society throughout the conquest of the frontcountry, their status as producers growing suspect within the context of their mimetic relations with wolves.

By the 1920s colonial authorities in Montana had articulated and institutionalized procedures for identifying and destroying predators—those humans and animals whose labors did not conform to the colonial economy's preferred representations of value. In Alberta the institutionalization of a predator-producer binary was more ambiguous. This chapter grounds the development of these bounty systems in the environmental, cultural, and labor histories of wolfing. It is organized into two main sections: a discussion of the gradual "animalization" and social marginalization of wolfers as they increased their mimetic engagement with predators, and a comparative analysis of the institutionalization of bounty systems in Montana and Alberta. Together, these discussions reveal not just a labor history of wolfing but also the history of labor as a concept in the frontcountry—how it emerged from relations between humans, animals, and colonial authorities in their various attempts to capture and represent value on the northwestern plains under concepts of predation and production.

Living Like a Wolfer

Wolfers' interactions with wolves were intimate and mimetic. To an extent, killing wolves required wolfers to "live like wolves," as Peter Koch described, but it also required the maintenance of human identity, a social problem that wolfers struggled to resolve from the 1860s through the early 1920s. The technologies of wolf bounties were more complicated than steel traps, strychnine, and .30-30s. They also included new modes of encounter, communication, and interpretation with wolves and other nonhuman beings. Like so many other physical metabolisms in the frontcountry, the wolfer's trade blurred production and predation. Predator eradication was less a function of a "war on wolves" as much as a broader set of political, cultural, and ecological interdependencies, all relied upon by colonial authorities who sought to transform the range, physically and culturally, into a site for the modern capitalist production of beef cattle.

While Koch managed to transcend his humble beginnings, most other wolfers, especially those who wolfed after the mid-1870s, had fewer opportunities. In the 1870s many people from diverse backgrounds did wolfing work. By the 1880s and 1890s its field of practitioners had narrowed to the poor and mostly unpropertied. By the early 1900s some bounty payments in Alberta were made only to Indians, and after another decade the work became federalized. Over time, wolfers drifted toward the margins of colonial society. These moves hinged on specific understandings of value and its origins, first on a trajectory of redefining labor as a solely human activity, and then on restricting the meaning of legitimate human labor to productive (and protective) rather than predatory forms of value creation. In the process, cattlemen elided their own predatory actions, as well as the productive actions of animals, wolfers, and others deemed "animal-like." Cattlemen animalized predators and humanized producers through their attempts to determine the form and practice of productive labor. The mimetic ability to "live like a wolf" was crucial to effectively killing wolves, but it was also a major obstacle to wolfers' abilities to claim their labor as legitimate.

In the 1860s and 1870s, when wolfers killed wolves for pelts rather than bounties, wolfing could hardly be described as either "hunting" or "trapping." The practice involved shooting a large animal such as a bison, an elk, or, later, an old beef bull, which wolfers then poisoned with strychnine. According to H. A. Riviere, each bait required about four ounces of the poison. After procuring a bait animal, wolfers rolled it onto its back and disemboweled it, slicing open the belly in a wide oval. With the organs and digestive tract removed, wolfers dumped strychnine granules into the bait animal's body cavity, called the "tub," where they slathered the poison with the animal's blood and chunks of its heart, liver, and other tissues. Then they scattered the blood and toxic morsels all around the body, killed another bait animal, and repeated the process. After spreading several of these baits over a wide swath of prairie, wolfers rode out to collect their kill every few days, sometimes returning with dozens of dead wolves. In the process of this poisoning, wolfers also killed hundreds of other plains scavengers—coyotes, foxes, magpies, cats, and dogs—in what has to be one of the most odious and destructive methods of "production" ever unleashed on the northern plains.[2]

Wolfers who baited and poisoned wolves held relations with wolves different from those who wolfed later and used different techniques. Unlike other methods of capturing wolves indirectly, like trapping them with leghold traps, baiting did not often require a great deal of care and attention to covering up one's human trace. When Koch maintained that these strychnine wolfers "lived like wolves," he meant so in more of a metaphorical than a metamorphic sense, one that developed increasingly from the 1890s onward as wolfers began to perfect new tactics of stalking and denning wolves. The actions of baiting and poisoning were undoubtedly dirty and dangerous. They also required a steady stomach and, likely, a general disdain for living creatures. These were all characteristics that immigrants like Koch would have associated with wolves.

The expansion of the beef industry onto the northern plains precipitated this major transformation. In 1883 the Montana Territorial Legislature passed a bounty law on wolves and coyotes at the behest of the Montana Stock Growers' Association that dramatically altered the scale and the practices of wolf killing. These laws lasted until the 1920s, when the U.S. Bureau of Biological Survey began federalizing the labors of predator eradication. Meanwhile, a host of other agencies and associations paid wolfers to kill wolves on public and private lands throughout Montana and Alberta. In Canada neither the Northwest Territories nor Alberta ever had a long-lasting provincial bounty bill in effect; however, the Western Stock Growers' Association, as well as dozens of large cattle outfits that leased Crown land, established their own bounty systems during the 1880s, many of which continued well into the 1960s. The last bounty bills in Montana were ended in 1962.[3]

During the bounty period, new wolfing practices emerged that helped legitimate this cultural transformation. Whereas before, wolfing had been a seasonal engagement, a temporary occupation for the upwardly mobile, by the mid-1880s this labor of extermination fell mostly to the poor, the unemployed, or the colonized. Animalized themselves as predators, wolfers stalked and killed wolves to reengineer the range for the needs of cattlemen, a reengineering that only further pushed wolfers toward the margins of colonial society.

At an 1888 meeting of the Montana Stock Growers' Association, T. C. Power reported that wolves killed about one-third of his calves each spring.[4] Clearly, his brisk two-decade-long business in wolf pelts, in which his firm alone shipped between ten and fifteen thousand pelts per year, had done little to control wolves and their numbers, who transitioned from eating skinned bison to livestock, an abundant source of predator-naive food. Although wolf depredations were certainly exaggerated out of political necessity, as Edward Curnow has argued, in the 1880s wolves certainly killed cattle in large enough numbers to be an economic and emotional concern for inexperienced cattlemen casting thousands of cattle out toward premature death at the eager fangs of lupine predators.[5]

Power, along with Pierre Wibaux and other MSGA members, continued to advocate for the use of poisons, even proposing that, in the absence of any more suitable game animals to use for bait, cowhands on spring and fall roundups be required to kill "all old bulls and worthless ponies . . . for the purpose of poisoning wolves and coyotes."[6] But by the 1880s poisoning was not as predictably efficient. This was likely a problem of predator naïveté. One Canadian wolfer, George Nelson, explained that "when the slaughter of livestock by wolves is at its peak, it is difficult to trap one of the marauders. They have plenty of fresh meat; they needn't prowl around old carcasses for their meals. Leave that to the coyotes!"[7] By the early 1880s, with the near annihilation of bison and the importation of hundreds of thousands of cattle, the frontcountry's predator-prey ecology had transformed so drastically that stockmen needed to develop new wolfing techniques.

Seeking to establish themselves as producers rather than predators, cattlemen generally sought to avoid visceral acts of killing. Wolfing was also a time-consuming activity and, for cattlemen, one better suited as an incidental project when riding on the range rather than a full-time occupation. Trapping and denning—without carbon monoxide—also required considerable time, along with skills and experience not easily mastered by amateurs. According to H. G. Pallister's history of the Alberta cattle industry, "ranchers had to take where the wolfers left off," but many ranchers themselves did not wolf, or at least did not do the dirty work of digging out dens or setting traps, preferring to simply shoot, and mostly miss, long-range targets of opportunity.[8]

Shooting wolves was generally ineffective. Learning to stay out of rifle range, wolves frustrated many ranchers by lurking hundreds of yards away. One Montana cowboy, T. B. Long, recalled that "it was very seldom that any of these wily killers were shot with a rifle. . . . They are just too smart to let one get close enough for a good shot. I have shot at a number of them, but I don't think I ever hit one."[9] Wolves' frequent nocturnal activity presented another challenge for would-be shooters. H. A. Riviere remembered a friend of his, Charley Vale, who ranched near the Porcupine Hills in southern Alberta and whose home ranch was often visited by wolves under cover of darkness. One night they chased his dogs into the house, who woke him up. Vale hustled out of bed with a loaded weapon. "He had picked up his rifle," according to Riviere, "and though he saw shadowy forms about the corrals, he also knew enough not to shoot at any unknown thing just because he thought it moved, so he just fired a few shots in the air and went back to bed."[10] Gerald Hughes, who grew up in the Judith Basin, recollected a similar experience to an interviewer who asked about Snowdrift, the infamous white wolf who killed dozens of cattle throughout the upper Missouri River valley and was one of the last wolves to be bountied in the state of Montana. Figuring out Snowdrift's general nighttime pattern, Hughes's father set up a blind on top of a hay rack near his herd. "Along midnight came the old white wolf," Hughes recalled. "Pa fired off several shots in the dark, but he couldn't see very good, so he didn't get him—just scared him good. But the bastard never come back."[11] And even under the best circumstances, when ranchers had a clean shot at a wolf, hitting one was never a sure thing. Running low along cover, wolves presented small moving targets seldom closer than a hundred yards, and it was almost impossible for average marksmen to consistently hit them until the introduction of flatter-shooting cartridges such as the .270 during the 1920s. Instead, firing heavier calibers with less accurate trajectories, wolfers and cattlemen from the 1880s through the 1920s needed considerable practice to hit wolves and coyotes from long distances.

More effective wolfing innovations were developed by an emerging class of full-time professionals. Stricken by a lack of suitable bait animals following the near annihilation of native ruminants, wolfers largely abandoned strychnine poisoning, which had served them well prior to the 1880s. In its place, wolfers began a combination of trapping, shooting, and denning wolves, all of which were time-consuming, sometimes frustrating work. Ewen Cameron, an early rancher and writer and the husband of the prominent photographer Evelyn Cameron, described one wolfer "named Maurice Barret, who follows on horseback any fresh-found wolf-track in the snow, never hesitating to camp on the trail at dark should no ranch house be available. The hunted animal, unable to baffle or shake off so relentless a pursuer, at last seeks the asylum of despair in a badland cave, whence it is relentlessly smoked out and shot."[12] According to Cameron, Barret was much like a wolf himself, a characteristic increasingly associated with wolfers toward the end of the nineteenth century. For the bulk of stock raisers, tenacity and adaptability, along with bloodthirstiness, represented the mimetic requirements of wolfers to their quarry.

Of all the tactics wolfers developed in the 1880s, denning, conducted during the spring, was usually the most effective. Wolfers located dens by following mother wolves gathering food for their pups. Tracking them back to their dens, wolfers had to seal the exits closed with dirt and then kill the den's occupants. Wolfers could accomplish this a number of ways, depending on whether any grown wolves were present. If an angry mother wolf was lying cornered in the den, then wolfers would take the time to dig out the den until they had a clear shot with a firearm. Otherwise, wolfers could often just crawl into the den and pull the pups out one by one. In the mountains, dens were difficult to locate, but out on the plains, wolves seem to have located their dens mostly in large cutbanks along rivers and streams. In these circumstances, denning was an extremely effective method of locally eradicating wolves, although it rarely accounted for adult wolves, and it was also likely that the breeding pair would birth a new litter or that new wolves would take over the pair's former range. Despite this limitation, denning seems to have caught on at least by the late 1890s, when the state of Montana began paying separate bounties for wolf pups at a lower rate than adult wolves.[13]

Besides just representing the wild and untamable, wolves provided frightening reminders of the ambiguity of human dominion over companion species relationships. Vivian Ellis, who grew up on ranches near both the Smith River in Montana and the Cypress Hills in Saskatchewan, recalled being scared of wolves as a youngster, even though she knew they rarely attacked people. "One night a wolf killed a colt," she remembered. "Dad had thought it would be safe enough that close to the house, but it wasn't. We heard the dogs barking and Dad got his gun and went out, but it was so black that night that you couldn't see anything, and all he could do was shoot into the air. No wonder I was so frightened."[14] Wolves prowled the fatal perimeters of domestic spaces not only killing cattle but also shattering any illusions colonists might have had that they were masters of their animal property.

This cultural activity was part of a wider attempt to dissociate the everyday violence of stock growing from its more protective, productive labors. Jon Coleman is right to point out that "in the course of becoming the most dominant predator on the continent, Euro-Americans often conceived of themselves as prey."[15] Throughout their colonial endeavors on the northern plains, settlers were quick to develop a defensive outlook, to view themselves as the ones under the siege of predatory violence rather than the ones perpetrating it. Custer's Last Stand is a good example of this colonial inversion of violence, as Richard Slotkin and other scholars have so thoroughly demonstrated.[16]

But more significantly, the livestock industry picked up this ethos of defense and used it to transfer ideologies and discourses of labor, reproduction, race, gender, and animality. In accordance with a notion of just defense and an understanding of the labor of livestock production as the labor of protection, a cult of masculinity developed in wolf killing, but one limited to the destruction of wolves in so-called self-defense. While wolfing for profit was predatory, animalistic, and ambiguously gendered, killing a wolf in the heat of defending one's animal property or even one's self or human family became an enigmatically masculine act.

Around 1900, for instance, six wolves allegedly attacked George Lane, foreman and future owner of the massive Bar U Ranch. Immortalized on canvas by Charlie Russell, Lane slew five of the wolves with six shots from his revolver. The tall tale, never entirely dispelled, grew into a foundational myth for the stock industry in southern Alberta. Depicting Lane as a selfless hero battling the predatory environment around him, the story, the painting, and now the sculpture, installed at the Bar U Ranch National Historic Site—and a matter of serious contemporary controversy between ranchers and animal rights activists in southern Alberta—all obscured the stock industry's predation on beef cattle by emphasizing instead its protection of livestock against other predators. Lane's encounter with the wolves was not merely the mimetic violence of two species of predator battling over prey, as Cameron interpreted the wolfer's hunt, but was instead an act of violence elevated, justified, and humanized in the name of self-defense. In this sense, it was the form of violent engagement, rather than its end result, that determined its value. This practice of engagement was a means to maintain the difference between predator and protector beyond the weak tools offered by metaphor and language, notions like predator and protector, which quickly broke apart in the context of defending livestock only to postpone their slaughter.[17]

As the nineteenth century merged into the twentieth, the form of labor rather than the function of labor came to dominate institutionalization of human-animal boundaries in the frontcountry. Commenting on the influence of Darwinism and the emerging acceptance of the materialist understanding of humans as animals in the early twentieth century, Jackson Lears has argued that "the new version of the ascent of man required distinction from the beasts on cultural rather than spiritual grounds: release from bent-backed toil; deflection of the gaze upward, away from the muck of biological existence."[18] This was certainly the case with the stock industry's attempts to dissociate itself from the mimetic violence of wolfing. Distinguishing man from beast hinged on delineating predation from production. The social stakes of "living like a wolf" increased dramatically from the 1880s through the 1920s. The necessities of colonial production required disenchanted human-animal interactions constituted by commodity relationships, but the effective cultivation of animal capital on the northwestern plains in both lupine and bovine forms still demanded the engagement of animals as persons. These contradictory modes of interaction threw a wrench in standard calculations of value. In the absence of unassailable visions of human-animal difference, the colonial state was forced to adjudicate these categories through the bounty program and its institutionalization of production and predation.

Institutionalizing the Bounty

Both north and south of the international boundary, bounties on "problem animals" like wolves, coyotes, mountain lions, lynx, bobcats, bears, squirrels, prairie dogs, and crows proliferated from the 1880s until the 1920s. But the burdens of these bounties fell hardest on wolves, the traditional archnemeses of Anglo-American cattlemen. By the mid-1920s wolfers had effectively eradicated wolves south of the forty-ninth parallel and in most of southern Alberta. Bounties on wolves were often higher than on other animals and often exceeded market prices for cattle and horses. In 1916 Walter Huckvale, one-time president of Canada's Western Stock Growers' Association (WSGA), privately offered an astonishing one-hundred-dollar bounty on wolves in southeastern Alberta. More reasonable WSGA members offered bounties in the neighborhood of thirty to fifty dollars.[19] The public bounties offered by state, provincial, and territorial authorities provided somewhat less incentive to kill wolves but were still lucrative. Over a forty-year period, Montana's vaunted bounty system fluctuated between one and eight dollars per wolf, plenty to account for the deaths of 111,545 wolves and 886,367 coyotes by the end of 1927.[20] Thanks to bounties like these, wolf killing blossomed into its own industry, an adjunct but necessary component of livestock production in the frontcountry.

The Montana Territorial Legislature passed its first wolf bounty law in 1883, offering one dollar for the full skin of each slain wolf presented to a county officer and vouched for by two witnesses. Along with wolves, the bounty law provided fifty cents for coyotes and eight dollars for condemned bears and mountain lions.[21] Montana's bounty law would have a long career despite suffering from fraud, inefficiency, and weak political backing. It was created during the so-called cowboy legislature alongside a slew of antirustling measures proposed during the 1883 session, and it arrived at the height of the cattlemen's power. Over the next forty years, their influence faded as mining and industrial farming transformed the state's political landscape. Just a scant few years after its establishment, the law came under siege, and stock growers reluctantly added squirrels and prairie dogs to Montana's bounty law in order to appease farmers who grumbled over the bill's neglect to address crop-raiding animals. The bounty laws were expensive, and in 1887, after bleeding out two-thirds of its entire annual budget to encourage the slaughter of native fauna, the territorial legislature repealed the laws altogether. Not until two years after statehood, in 1891, did another workable bounty law emerge, but this state law lived well into the twentieth century.

North of the border, the Northwest Territories, in partnership with the WSGA, established a wolf bounty law in 1897. This program was a political disaster, and after it ran out of funds in 1902, it was never revived by Alberta's provincial government. Even less politically connected to Edmonton than Montana cattlemen were to Helena, southern Alberta's stock growers lobbied year after year for a new bounty law without success. Intent on developing most of Alberta's prairies and parklands into grain farms, the early provincial government found little political advantage in helping southern cattlemen with public money, particularly since so many stockmen ran herds on land leased from the Crown and had reputations as either British remittance men or rogue cowboys and ex–whiskey traders from the States.[22]

But despite a lack of government assistance, Albertans, like Montanans, also killed their fair share of wolves under private bounty arrangements. Due to a paucity of records, coming up with exact numbers of wolves killed under these private arrangements is impossible, but a variety of sources reveal an extensive system of labor created by public and private bounties. In short, in the Whoop-Up Country, wolfing, like stock raising, was an industry in its own right.

The transformation of wolf bounties in Montana from a desirable program for ranchers into an expected function of government worked out in more subtle ways as well. Even the seemingly simple task of defining what constituted a wolf became the necessary role of the state, and Montana's transition to statehood streamlined the exercise of this power. In 1893, two years after the state legislature enacted its first bounty law, it revised its payments to award five dollars for an adult wolf but only two dollars for a pup.[23] Since nobody actually specified the difference between an adult and a pup, this ruling created a bureaucratic nightmare. Not until 1899 did the legislature revise the law to state the difference in unequivocal terms: pups were pups only until November 1; after that they counted as adults.[24] But this was still problematic, since wolfers could simply den out pups and keep them alive until after November 1 to earn a 150 percent raise on their bounty claim. Finally, in 1913, the state examiner stipulated a more sensible ruling, defining an adult wolf by standardized measurements of its length, height, ear, nose, and head.[25]

The state of Montana also faced the necessity of developing a fraud-proof system of verifying that bountied animals had not already been bountied before. In 1909 the Bureau of Biological Survey circulated a pamphlet to state authorities advising them of the multitude of fraudulent possibilities for trafficking dead predators across state lines to receive multiple bounties. Written by the noted government wolf hunter Vernon Bailey, the circular argued that "the common practice of paying bounty on scalps alone, or in some cases merely the ears, is dangerous as even an expert can not always positively identify such fragments."[26] Montana did not really have this problem, as its bounty laws required claimants to present the entire skin of the animal for inspection to a county inspector, who would punch the skin and return it to the claimant. But later in the 1910s, Montana revised its policy to require only the heads of wolves and coyotes for bounty certificates, which would remain in the possession of county authorities and be destroyed monthly.[27]

Montana's bounty laws were also never very well funded, which led to unexpected and complicated problems. Government appropriations funded both the territorial bounty and the original state bounty, but in 1899 the legislature established a special annual tax assessed on livestock in order to fund bounty claims. This tax was negligible, averaging around four mills per head, or four-tenths of one cent; a large stock grower with 2,500 cattle would pay about ten dollars, only enough to fund the bounties on two wolves but significantly cheaper than hiring a professional wolfer and paying him wages. This bounty fund provided an enormous subsidy to ranchers and wolfers. In 1899 alone, wolfers bountied 23,575 wolves, which would have required assessments on over 30 million head of cattle.[28] The annual shortfall usually left the bounty fund around $100,000 behind, which amounted to almost a four-year delay for the payment of bounty claims. The State Treasurer's Office only received the assessment money once a year and did its best to apply it as far as possible.[29]

This budgetary shortfall was an obvious source of discontent for wolfers, who sent hundreds of letters to the Montana State Board of Examiners demanding payment. "I would like to call your attention to my Bounty Claim," wrote one claimant named Sam Davis. "Its about time I received it. It is over 4 years since I sent it to you!" In response to these queries, the Board of Examiners tried to remain as good-humored as possible: "Relative to your Bounty Claim, I beg to advise you that same will not be reached for payment for about one year yet. Am sorry that I cannot at this time aid you in the slaughter of more coyotes."[30] This five-year delay in payments provided a nice subsidy for the stock industry, which was not required to substantially increase its assessments in order to pay the bounty fund forward.

Instead, local banks and other speculators picked up the slack by cashing many wolfers' bounty certificates for varying cuts of the payments. Guaranteed by the state of Montana, these were safe debts to purchase, and it appears that no claims were left unpaid, despite the lengthy delays. But risk on the bounty certificates would have been determined by the unpredictable length of the repayment cycle, which was in turn driven by annual revenue from the livestock assessment, the number of predatory animals bountied, and the bounty rates on different predators, which tended to change every few years and generally increased above the rate of inflation until the 1920s. From 1901 to 1914, for instance, the bounty rate on adult wolves rose threefold, from five to fifteen dollars.[31] During the same period, the livestock assessment remained comparatively steady, but the number of wolves bountied actually decreased by around two-thirds.[32] All these considerations might have complicated bankers' fee calculations on the bounty claims. Then again, most wolfers needed their money sooner than the government could pay and grudgingly paid any reasonable fee to the bankers in order to get their cash.

This transformation of wolves into money relied on a partnership of colonial government with private finance. Montana's bounty system presumably existed as a corollary to the livestock industry's reproduction of capital through animal bodies. Both killing wolves and providing incentives for others to kill wolves was, by the early twentieth century, increasingly understood to be a service rendered by the government, at federal and state levels, for the benefit of Montana stock growers who sought to monopolize their own productions and representations of animal capital: beef cattle consumed by the stockyards of Chicago and Kansas City, not cows and calves hamstrung and consumed by the northwestern plains' nonhuman predators. But the state of Montana could not assure and administer this result without also providing the basis for another economy of animal capital: the capture and slaughter of predatory animals for bounty, greased and financed by the state's private banks. Wolfing developed into a profitable business, too profitable, in fact, to wholly serve the interests of stock growers. For wolfers, along with the bankers who served them, a permanent "war on wolves" was preferable to wolves' wholesale extermination.

Under these circumstances, "fraud" was fairly prevalent in Montana's bounty system. In many instances, fraud cases were a matter of contested definitions. There were always gray areas where wolfers, bankers, and others tried to squeeze extra booty from the wolf bounty. In 1918, for instance, a northern Montana merchant named T. J. Troy was trading for wolf and coyote pelts with the Cree and Blackfoot on the Rocky Boy's Reservation and then submitting bounties on the furs. Of course, this was illegal; the law required all bounty claimants, along with a county bounty inspector and two witnesses, to certify that they had personally killed the animals. The Board of Examiners prosecuted Troy for fraud, but he got off the hook thanks to endorsements by his friends E. J. Broadwater and Walter Brown, both prominent stockmen from powerful families in Montana state politics. Broadwater and Brown wrote letters to the state livestock commissioner and asked for his support in Troy's case, arguing that Troy merely misunderstood the procedure for transferring bounty certificates and that he really intended to submit the claims on behalf of his Indian clients. "I would appreciate it very much," wrote Brown, "if you would let Mr. Troy's claims go as he is doing good work in helping the Indians to catch those animals."[33] Fortunately for Troy, this turned out to be a suitable defense.

The most common of these frauds was to engage in a sort of "wolf ranching," either raising pups in captivity until they were large enough to bounty as adults or simply finding dens and annually killing only the pups rather than the den's pair of breeding adults. Ewen Cameron considered this strategy part of a wolfer's illicit existence as a predatory animal. Wolfers' mimetic abilities extended dangerously to this "well known trick," he explained, "of liberating she-wolves to multiply for the bounty."[34] According to the *Helena Daily Herald*, even county commissioners got a piece of this action:

> The late lamented Legislature passed a bounty law fixing the price to be paid for the scalps of wolves and coyotes at $3 each. Since the enactment of the law a number of cowboys have resigned from the range and gone into the bounty business for there is more in it. But these gentlemen are discounted by Commissioner Barton of Choteau County, who, according to the River Press, has purchased from a Cree Indian a she-wolf and litter of nineteen pups that were captured alive. According to the River Press, "it is the intention of Mr. Barton to kill the animals and secure

a bounty on the scalps." The deal will be a profitable one no doubt and could be made still more profitable if the thrifty commissioner would kill only the pups, secure a mate for the she-wolf and go into the breeding business. If she is as prolific at all times as with the litter in question there is money in it.[35]

In Alberta, H. A. Riviere recalled a similar situation with the Nakoda: "When the wolves were doing a lot of damage, some of the big outfits put bounties on wolves. The Stoney Indians knew all the dens, and raided them yearly, but took great care to leave the old ones unmolested. They figured that 5 or 6 cubs at 10 dollars a head, beat $10 for 1 female."[36] As Riviere's reminiscence suggests, authorities were fairly powerless to prevent the development of this shadow livestock industry in which wolfers raised predators merely to bring them in for bounty payments. And when Nakoda wolfers did this, as Riviere alleged, not only did they earn good money, but they created a source of animal capital both within and outside the boundaries of the colonial borderland's political ecology.

In contrast to Montana, the institutionalization of bounty laws in Alberta never really occurred due in part to a lack of government support but also due to an ineffective bounty infrastructure set up by Canada's Western Stock Growers' Association. For generations, Canadians have experienced the complaints and frustrations of ranchers in the prairie provinces grieving over their lack of support from Edmonton, Regina, Winnipeg, and Ottawa. A robust historiography of stock growing in the Canadian West underpins this populist sentiment, contending that, unlike in the U.S. West, Canadian governments stood in the way of ranching interests, favoring the claims to the land of Indians, farmers, miners, or environmentalists.[37] Recently, historians such as Warren Elofson have countered this thesis, arguing that crime, inexperience, environmental hazards, and unfavorable markets rather than political indifference stunted western Canada's cattle boom.[38] While Elofson's contentions are accurate and well documented and provide a landmark intervention in the history of Canadian stock growing, neither his nor previous narratives adequately explain the failure of Alberta's cattlemen to eradicate wolves. It is undoubtedly true that Alberta's remote wild places helped shelter wolves from dogs, traps, and bullets. Likewise, an impermanent and underfunded system of private and public bounties provided less incentive for full-time wolfers in Alberta than in Montana. But the failure to solidify bounty institutions in Alberta developed alongside Canadian stock growers' relative failure to institutionalize their own representations of labor and value as the dominant visions of production north of the forty-ninth parallel. As such, Canadian stock growers could not sufficiently transform their states' colonial prerogatives to ones synonymous with their own.

One of the Alberta bounty system's main problems was its lack of central administration. Prior to the 1896 establishment of the WSGA, stock growers on the Canadian plains were not organized in any formal association that pressed for uniform bounty legislation across the territory. Instead, large stock-growing families like the Cochranes, Hatfields, and others employed their own full-time wolfers or placed private bounties on wolves and coyotes killed on their own property. Although some of these bounties were quite large, which made wolfing a lucrative proposition, the inconsistency and impermanence of these private bounties crippled efforts to control wolf populations. Hired wolfers were effective at controlling local populations of predators at the home ranches and perhaps also in open-range country during spring and fall roundups, but their geographical confinement limited the extensiveness of wolfing operations compared to those south of the border. In 1897 one of the WSGA's first unified actions was to place a public bounty on wolves, ten dollars on adults and two dollars on pups. Even so, unlike Montana's well-developed bounty laws, the WSGA's bounty agreement did not include any specific procedures to streamline the effectiveness of bounty hunting. Rather than designating traveling bounty inspectors by county or district, as was the custom in Montana, the WSGA required wolfers to travel to one of three locations in the territory, either Calgary, Macleod, or Maple Creek, to cash their claim.[39] This probably streamlined WSGA paperwork, but it handicapped the establishment of a standard regional bounty system. Rather than wolfing for the WSGA bounty, which was less than most private bounties and could require traveling up to three hundred miles, wolfers continued to work for independent cattlemen and their bounties. The WSGA recognized this problem, but over the next few years, several motions to expand their bounty operations to Pincher Creek and Medicine Hat failed.[40]

The WSGA bounty was also chronically underfunded, and no financial apparatus developed to pick up the slack, as it had in Montana. The 1897 bounty had been instated with only $1,000 and was to be funded only by WSGA member dues.[41] Sometime shortly thereafter, the territorial government agreed to match the bounty fund, and in 1900 the WSGA raised its bounty rates to fifteen dollars on adults and five dollars on pups.[42] This generous payment increase was the beginning of the end for the WSGA's bounty operation. In 1902 the association wrote the territorial government expressing that "the wolf bounty appropriation was not sufficient to meet the claims during the year" and requested a larger territorial appropriation.[43] A year earlier, the WSGA also suggested that 25 percent of revenue from Dominion grazing leases in the Northwest Territories be returned to ranchers in the form of a permanent appropriation to the wolf bounty fund.[44] With neither the territorial nor Dominion governments cooperating with the WSGA's demands, the bounty fund went insolvent. By May 1908 the fund had long been exhausted and the bounty operations suspended. Even worse, the WSGA had only $50.46 in its entire association accounts. The best it could do was petition the new provincial government of Alberta to reinstate the bounty, which it did without success.[45]

The WSGA bounty was also crippled by what seemed to be an increasingly lack of interest in wolfing above the forty-ninth parallel. While the Montana bounty helped create an unequivocal class of wolfers who worked at their occupations full time, the WSGA's poor incentives did not generate a sustained ambition to kill wolves and other predators. In fact, it appears that many of the WSGA's bounties were claimed by its own members. In 1900, after placing a bounty on coyotes in addition to wolves, the WSGA went so far as to ban members from collecting any more bounties in an effort partly to encourage others to take up the task and to avoid bankruptcy by paying fewer claims. Members' "interest in dead coyotes should be sufficient reward for them to kill coyotes at all opportunities," the decision stated. Later on in the meeting, the association passed another motion to pay the bounty only "to Indians and Half breeds."[46]

For these reasons, among others, the Alberta bounty's failure was linked to the wider political context of the province and its institutionalization of a colonial vision of production different from ranching: dryland farming. By the time the province had been established in 1905, dryland agriculture had emerged as the dominant paradigm in Canada's western politics, and the open-range cattlemen, with their vast, thirty-year Crown leases, represented a recognizable impediment to the ordered colonization and settlement of the wild rose country. While the cattle industry, centered in Macleod, continued to dominate the southern third of the province, the seat of state power resided nearly four hundred miles north, in metropolitan Edmonton, surrounded by farmers in and around the wetter Saskatchewan valley. Under these circumstances, even well-intentioned efforts to assist stock growers by the provincial government missed their mark. After intense lobbying by the WSGA, finally, around 1913, the Alberta Department of the Interior appropriated $3,000 to reestablish the wolf bounty. But this program turned out to be short-lived and of virtually no benefit to southern stock growers. According to Walter Huckvale, "Owing to the hardupness of nearly all farmers in the early part of last year a lot of them had gone out in the North and had killed so many wolves that they exhausted the bounty. We in the South never got a piece of it." This was an explanation confirmed by Duncan Marshall, the Alberta secretary of the interior, who was unable to lobby for additional money to cover the southern portion of the province.[47]

WSGA members did their best to drum up interest in the bounty with other organizations, but to little avail. In 1915 George Lane chaired a meeting in which the association mulled over arguments it could use to convince small stock owners and mixed farmers—by no means its traditional members—to join the association, pay membership dues, and support another wolf bounty. They developed a twofold pitch. First, they expressed "the necessity of keeping wolves down." Huckvale and others claimed that wolf depredations hit small stock growers especially hard, since even minor losses to their stock animals could utterly destroy their business. But small growers and mixed farmers who tended their animals close to their farms in feedlots and pastures did not face the same trouble with wolves as large operators who let their herds run amok across the range.⁴⁸ The needs of smallholders were rather different. As one WSGA member put it, describing the incompatibility of their association with the desires of others, "We have to demonstrate that we are some value as an association, look after transportation facilities and concern ourselves with the interests of the producer, otherwise he is not going to follow or join us."⁴⁹ In Alberta the cattlemen had failed to wholly institutionalize themselves as producers, partly through their failure to identify and objectify predators. In one of their final requests to reinstate a government wolf bounty, the best the WSGA could muster was a feeble assertion that wolf eradication was a public good; reestablishing the bounty, they argued, would "[relieve] to an extent, the public spirited men who have done so much for the general public in this way."⁵⁰

The institutionalization of bounty laws developed unevenly across the Montana-Alberta borderland. While the state of Montana created a long-term system, Albertans struggled to replicate this system in the North. Undoubtedly, wolfing served the interests of livestock producers, but practices like wolf farming also held subversive possibilities. By opening an entrepreneurial space for indigenous people, poor whites, and others at the margins of colonial society, wolfing provided an alternative to outright proletarianization. And the scale of its state- and corporate-sponsored violence reflected ambivalently on the colonial livestock regime, begging the question, If stock producers weren't also predators, then why not? In Montana, the state bounty system settled this question by identifying predators and casting them outside the boundaries of legitimate production. In Alberta, this question lasted as a more ambiguous problem.

Conclusion

In a strictly economic sense, wolf bounties did not provide the livestock industry with the returns it expected. In Montana, forty years of bounty laws had helped reduce wolf populations to a nearly imperceptible level; however, this did not save the state's cattle industry from periodic cycles of crisis, including the postwar bust in 1919 and 1920 that all but demolished the state's large operators. Likewise, Alberta's extensive network of private bounty arrangements failed to avert the devastation of Canada's open-range cattle industry. Although the dominant form of stock production until the early twentieth century, large-scale ranching was already obsolete in Alberta by the 1890s; the 1910s witnessed its death knell.[51] By this time, with or without government support, open-range cattle ranching in the frontcountry faced substantial headwinds. With more and more land fenced, homesteaded, or under private ownership, successfully grazing cattle required more expense and attention than ever before. In addition, new taxes and fees accompanied the leasing of federal, state, and provincial lands for grazing purposes, and running cattle on these typically higher-elevation refuges brought livestock into closer proximity with the few dozens of wolves who survived the bounty programs, along with other stock-killing predators such as mountain lions and grizzly bears. In many respects, the bounty laws so vehemently supported by the borderland's large stock growers were an abject failure. With great expense and effort, stock growers managed to remove wolves from vast portions of prime grazing land, only to lose much of that land to the termination of Crown leases, an influx of dry farmers, and numerous bankruptcies. And in the process, stock growers established regulatory bureaucracies that many cattlemen, both north and south of the border, would soon identify as major obstacles blocking the success of their own operations.

In this sense, predator control was more than just a matter of protecting livestock, it was also a battle to represent value in a manner that justified the violence and exploitation contained within the colonial open-range cattle industry. Wolfing was a means to "add value" to the metabolic relations of cattle and cattlemen without necessarily creating any tangible economic gains for cattlemen. What the human-animal relations of wolfing generated instead were new understandings of labor and value that naturalized the folkways and political ecologies of the stock industry. Both north and south of the forty-ninth parallel, stockmen mobilized the overwhelming cultural acceptance of their images and behaviors to frame predator control as a debate within the business rather than a debate concerning wider circulations of value, that value's origins, and more democratic alternatives to its distribution. If only ambiguously successful economically, wolf bounty programs proved decisive in rendering alternative representations of value as predatory and illegitimate—outside the scope of the colonial frontcountry's new capitalist boundaries.

An important fur-bearing animal during the 1870s and then a valuable beast of bounty under the predator control laws of the late nineteenth and early twentieth centuries, wolves have been significant economic actors throughout Montana's history. Before the 1880s, an array of economic relationships with wolves and other animals that were focused on hunting rather than livestock growing had provided many Montanans with alternatives to cattle and sheep raising. But by the 1900s the institutionalization of the bounty system had helped establish a far different set of economic and cultural expectations regarding cattle, wolves, and other animals. Ranchers and domesticated livestock signified a productive landscape deserving of state and federal support, while wolves, hunters, and others who killed large animals came to represent Montana's predatory past, a colonial history that the region's emerging elites were eager to disavow.

Given the monetary incentive to produce predators for state bounty payments, it was never in the best interest of wolfers to eradicate wolves from Montana. In fact, it was salaried hunters working through the U.S. Bureau of Biological Survey who accomplished the feat during the 1920s and 1930s, focusing their efforts on federal lands that had become, in many cases, high-altitude refuges for the carnivores. Around this same moment, hunting itself increasingly transformed into a recreational activity of the privileged rather than a means of earning a living, as the next chapter explores. By the early twentieth century, the killing of Northern Rockies wolves had helped sharpen popular distinctions between productive and predatory labor that celebrated stock raising as productive work and denigrated hunting—including wolf hunting—as illegitimate labor within the new economy.

3

Making Meat

When bison began to disappear from the plains of northern Montana in the fall of 1883, life on the Blackfeet Reservation entered an era of destitution. By the next spring's thaw, perhaps one in three Blackfeet on the reservation had starved to death.[1] Constrained within their reservation boundaries and no longer able to produce their own subsistence by hunting, the Blackfeet would soon have to sell their labor for government rations—mostly beef. History seemed to collapse under this transformation in Blackfeet food and work. Blackfeet elders stopped recording events in their winter counts, with the exception of listing the names of their many friends who died. One man named Elk-horn expressed his own lack of interest in tracking the passing decades: "Since our people were confined to the limits of the reservation, nothing else has happened worth remembering."[2] Many observers believed, incorrectly, that the Blackfeet and their way of life were vanishing. But throughout the nineteenth century's final decades Indian Office administrators dealt daily with recurrences of the Blackfeet's "lost traditions," and few were more troubling than Blackfeet ways of making meat.

The origin, preparation, and consumption of meat provided constant reminders of the problems that both plagued the reservation and accompanied industrial capital's expansion across the northern plains. On the Blackfeet Reservation, meat provided a crucial means by which the Office of Indian Affairs (OIA) could work toward subordinating Blackfeet land and labor—two major goals of federal assimilation policy. By regulating access to meat, the OIA believed it could transform the Blackfeet from hunters to herders, thereby confining the Blackfeet within their reservation boundaries and restructuring Blackfeet patterns of subsistence to conform with the institution of wage labor and capitalism in northern Montana. For the Blackfeet, the message was clear: they needed to give up hunting, and they needed to either "work or starve."[3]

But colonizing the Blackfeet through their stomachs was a difficult undertaking, primarily because the Blackfeet refused to give up their old traditions of meat making. As government-issued beef cattle replaced the reservation's dwindling herds of bison, the Blackfeet continued to make meat in much the same manner: by killing it when they were hungry. Moreover, they ate it raw, smoked, and sometimes baked whole in the earth. Civilizing the Blackfeet would require more than the mere replacement of one indigenous ungulate for a domestic Anglo-American animal, it would also require a revolution in the Blackfeet's practices of killing and consuming.

The OIA enforced this transformation through the construction of an agency slaughterhouse, a building born from the administration's obsession with producing "clean meat." The filth associated with traditional Blackfeet methods of slaughter aroused the compulsions of OIA agents and inspectors. For these colonial personnel, however, "clean meat" was more than simply sanitary meat; it was meat that—like the Blackfeet themselves—had to be purged of its predatory history. By providing a location to centralize and regulate the killing of beef cattle behind closed doors, the first agency slaughterhouse offered administrators an effective way to dissociate this meat from its animal origins, a crucial step in cleansing its predaceous past. Furthermore, by representing traditional Blackfeet meat-making labor as predatory rather than productive, agency administrators increased their discipline over those Blackfeet who continued to kill and eat cattle in the field. This first agency slaughterhouse provided both the political and the physical means to subordinate Blackfeet land and labor within Montana's growing livestock industry.

Transformations in Blackfeet meat preparation and consumption, however, were not fully expressed until the construction of a second agency slaughterhouse in 1895, the year the OIA moved the Blackfeet Agency from its old location on Badger Creek to its present location in Browning. Designed to address the sanitary shortcomings of the Badger Creek slaughterhouse, the Browning Agency featured both a slaughterhouse and a butcher's shop where the distribution of cattle and their edible components could be further refined and controlled. While the first slaughterhouse provided a key institution in delegitimating the Blackfeet's traditional male labors of hunting, the second slaughterhouse and its accompanying butcher shop worked to delegitimate traditionally female labors of butchering and meat preparation. In the process, the OIA developed more complete governance not only over meat's production but also over its consumption. By transforming Blackfeet foodways, the construction of these two agency slaughterhouses provided a potent force to colonize Blackfeet land and labor during the late nineteenth century.

Bison and Blackfeet Foodways

Bison meat dominated the Blackfeet diet until the introduction of beef cattle to the reservation in 1879. Blackfeet kinship practices that linked humans to wolves as bison hunters also inflected indigenous food traditions, as discussed previously. Even hardened Indian agents commented with disbelief on the Blackfeet's wolfishness, one claiming, for instance, that the Blackfeet were "nearer to barbarians than anything I have ever seen."[4] The Blackfeet's heavy diet of animal flesh played a large role in these barbaric estimations. Other than seasonal harvests of berries and wild turnips, in good times the Blackfeet ate meat and nearly always that of mammals. Before the introduction of horses and rifles in the late eighteenth century, Blackfoot hunters chased bison on foot, cooperatively driving small herds of the animals over cliffs or sometimes into corrals, where they could be speared. Ethnohistorical and archaeological evidence both suggest that these Blackfoot hunters managed the location of bison herds through a practice of rotationally burning prairie grasses. The Blackfoot did not merely wander the plains following bison, as historians have sometimes misunderstood, but instead kept the animals close by managing certain locales—particularly those near *pisskan*s, or "buffalo jumps"—as optimal bison habitats.⁵

Rather than "cooking" meat in the usual Anglo-American sense, the Blackfeet relied instead on the northern plains' abundant wind and sun to prepare bison flesh for consumption, most often slicing meat into thin strips for drying. In addition to eating bison's lean flesh, the Blackfeet developed sophisticated techniques that rendered the animal's blood, tallow, bones, and entrails into edible and preservable foods. *Dupuyer*, for instance, best described as a kind of bacon made from the smoked fat of bison, quickly captured the palettes of French and Anglo-American fur traders, who eagerly acquired the food from the Blackfeet as a luxury substitute for the traders' usual diet of pemmican. Making dupuyer was a fairly simple process: Blackfeet women—the traditional butchers—hung large chunks of otherwise inedible tallow from the poles of their lodges, allowing it to slowly smolder above the fire for a period of days to weeks. This process produced a lightweight, long-lasting food that Blackfeet war parties carried on distant campaigns against their Crow and Cree enemies. The Blackfeet also produced a kind of giant blood sausage, packing the large intestine with pooled blood, fat, berries, and other ingredients and then slowly roasting it in a bath of hot ashes. Another Blackfeet delicacy was a fetal calf removed from its slaughtered mother and roasted whole in an earthen pit lined and covered with ashes and hot rocks. By the 1880s these techniques were easily translated from the bodies of bison to those of beef cattle, disgusting those white observers grown used to those late nineteenth-century standards of civilization—roasts and beefsteaks.⁶

In addition to eating bison, the Blackfeet utilized bison furs as an important trade item from which they procured horses and rifles, as well as food goods such as tea, coffee, sugar, flour, and whiskey. These robes had become significant commodities in the northern plains fur trade by the mid-nineteenth century, as demand grew for luxurious pelts to upholster Victorian sleighs and carriages. Bison robes differed from bison hides in that they were more than just tanned, leathery skins; they were intact pelts that included all the bison's hair. Supply for these robes was limited to bison herds on the northern half of the animals' range, where nineteenth-century observers claimed that their hair grew longer and thicker because of the winters' subzero temperatures. The skilled labor required for making these robes also limited their supply. In the case of the Blackfoot, women were responsible for this labor, and they spent hours stripping flesh and tissue from the robes and rendering them into supple furs by smearing them with brains. In a detailed study of Cheyenne robe production, anthropologist John Moore has estimated that through this unpaid labor, Cheyennes alone put $50 million of capital into the hands of white merchants by the Civil War.[7] The scale of the Blackfoot trade was even larger, and by the 1860s Fort Benton, located on the highest steamboat-navigable reach of the Missouri River, had blossomed into one of Montana Territory's most important towns, primarily as the entrepôt of this robe trade. Between 1859 and 1884, when some of the last bison carcasses floated downriver, traders in this "world's innermost port" had shipped out over 760,000 bison robes, nearly all of which were produced by Blackfoot men and women.[8]

Other research has noted that participation in the bison-robe market altered Blackfoot subsistence and food relationships in more subtle ways. Like other Plains people—the Arapaho, Crow, and Lakota, for instance—the Blackfoot's martial culture was driven by the introduction of horses and rifles in the late eighteenth century. In the case of the Blackfoot, horses became a form of currency for fulfilling social obligations. Not only were horses essential tools of war, but often they were also war's object, and successful warriors accumulated fortunes in the animals that they leveraged for political power within their extra-familial bands. As research by David Nugent has demonstrated, such wealth in horses altered Blackfoot social hierarchies in the early nineteenth century, allowing for the emergence of smaller-sized bands that were dominated by one or two extremely wealthy band chiefs, and on whom most other band members were economically dependent. The distribution of meat from the hunt, however, while centralized under the control of band chiefs, remained largely egalitarian.[9]

If an uneven trade that destroyed the Blackfeet's traditional food source was not enough, a series of American and Canadian military campaigns during the 1870s forced the Blackfeet to restrict their movements and negotiate away their land base. On the morning of January 23, 1870, the U.S. Second

Cavalry Regiment took firing positions on a bluff above the Marias River, a three-day ride into the heart of Montana's Blackfeet Reservation. Below the soldiers, along a bend gouged by the river's frozen flow of ice and mud, sprawled a Pikuni winter camp sheltering several hundred people from the brutal cold. Earlier that week, a dome of arctic air had spilled southward, dropping temperatures into the minus thirties. The troops had marched through the cold from Fort Shaw, on the reservation's southern boundary, where most of the force had arrived just days earlier from Fort Ellis, near the town of Bozeman, over three hundred miles to the southeast. Summoned by orders from General Phil Sheridan, the Second Cavalry had journeyed across Montana undetected, traveling by night, resting by day, lighting no campfires. To boost morale and to provide an illusion of warmth, Major Eugene Baker, their commanding officer, allegedly dispensed whiskey rations prior to their assault.[10]

According to Baker's official report, his regiment took the camp by surprise and pounded it with rifle fire for a solid hour, shooting everything that moved. The regiment's first American casualty of the day was a trooper thrown from his horse as the regiment charged down the steep bluff and into the Pikuni camp, pulling down lodges and killing survivors. Its second casualty was a soldier shot in the face as he entered a lodge to murder its occupants. By 11:00 a.m., Baker's surprise attack had killed 173 Pikunis and captured over 100 prisoners, with the loss of two soldiers. Leaving behind one company of troops to guard the prisoners and to finish obliterating the camp—literally piling together all Blackfeet possessions and setting them ablaze—Baker's regiment rode ten miles downriver to attack another band of Pikunis, one led by Mountain Chief, Baker's original target. He arrived to find Mountain Chief's camp hastily evacuated. But rather than continue his pursuit, Baker led his troops back to Fort Shaw, satisfied with his indiscriminate assault, lauded by his lieutenant as "the greatest slaughter of Indians ever made by U.S. troops."[11] An investigation and eye-witness reports confirmed that Major Baker and most of his troops were drunk. Earlier that winter, medical records from Baker's command at Fort Shaw indicated that between 6 and 12 percent of his soldiers were unfit for duty as a result of consistent drunkenness and severe hangovers. Baker himself was court-martialed two years later after the so-called Battle of Poker Flats, a fight on the upper Yellowstone that he missed while drinking and playing cards in a tent with his lieutenants.[12]

The Blackfoot's situation was similarly grim north of the international boundary. Although no representatives of the Canadian government committed genocidal violence on the same scale as the U.S. Army, the North-West Mounted Police nevertheless subjugated and confined the Blackfoot onto reservations, in turn opening up land for white settlement and limiting Blackfoot opportunities for off-reservation hunting. In 1877 the Canadian government forced the Blackfoot to sign Treaty Seven, confining them to three reserves in southern Alberta, subduing the Blackfoot "by implementing Ottawa's policy of coerced assimilation," as the historian Andrew Graybill has described.[13] By the late 1870s these colonial pressures, together with the rapid destruction of the bison, had unsettled Blackfoot foodways and left the Blackfeet on the brink of a subsistence crisis.

Beef Assimilation and the First Reservation Slaughterhouse

In 1884 the OIA built the Blackfeet Reservation's first slaughterhouse, a central element of its civilizing strategy because of the way it confined and categorized the labor associated with killing animals to make meat. Located at the Badger Creek Agency, the facility measured six hundred square feet, with a low ceiling and drafty walls built of hewn logs. By enclosing the production of meat in such a space subject to the supervision of Indian Office administrators, the agency used the slaughterhouse to extend its control over the definition of legitimate labor on the reservation, a definition that no longer encompassed hunting. The OIA hoped that, for the average Blackfeet, making meat would become a process of exchange rather than a direct human-animal interaction, thereby conforming to Anglo-American standards of wages and market consumption. This assimilation project also corresponded with the broader expansion of capitalist stock raising on the northern plains. Transforming the Blackfeet from hunters to herders would create a new market for cattle while also providing the rationale to confine the Blackfeet to a smaller reservation. In its subordination of Blackfeet labor, the agency slaughterhouse both reflected and enabled the pursuit of these goals.

At first, confining the Blackfeet came at a cost to the U.S. government. As part of the Blackfeet's treaty negotiations, the tribe had agreed to open their former lands south of the Missouri River to white settlement in 1865 in exchange for $50,000 worth of annuity goods distributed over the next twenty years.[14] The exact kinds of goods were not specifically defined, but they originally included flour, coffee, and sugar, along with items of varying uselessness to the Blackfeet, like fishhooks.[15] Although these goods were originally distributed to all enrolled members of the tribe, in 1878 the OIA mandated the bureau-wide adherence to a new policy of making Indians exchange their labor for rations tickets, which could then be used by individuals to purchase annuity goods. By seeking to make the Blackfeet liable for their own subsistence through this form of semipaid labor, the order underscored the OIA's goal to assimilate the Blackfeet, as well as its obligation to cut its expenditures, which had soared since the establishment of President Grant's "peace policy." Grant's strategy to avert violence between white settlers and Indians hinged on the prohibition of off-reservation hunting, which only increased Native reliance on food rations.[16]

The OIA envisioned that eligible labor for Blackfeet rations tickets would include agricultural work performed at the agency farm, a pathetic field of frost-bitten vegetables and rotten potatoes. But the Blackfeet's agent, John Young, protested the order on the basis of its infeasibility, ultimately winning the reservation's exemption from the commissioner of Indian Affairs. Not only was the cultivation of garden crops hopeless along most of the northern frontcountry, claimed Young, but most of his Indians simply refused to work.[17] Young suggested instead the establishment of an agency cattle herd from which he could build a reservation stock industry modeled on the vast open-range cattle operations that had recently enveloped the rest of north-central Montana. According to Young's plan, the Blackfeet were to become cattlemen—they were to live off beef, and not just its calories but also its cash proceeds.

In 1879 the Office of Indian Affairs accepted Young's plan and issued the agency its first delivery of fifty-six animals, augmented with an additional herd of five hundred cattle in 1880.[18] The primary purpose of this herd was to develop a long-term stock industry on the reservation, not to serve as an emergency source of subsistence. Young estimated it would take until roughly 1885 before the herd grew large enough to sustain the Blackfeet's beef demands. Until that date, Young attempted to keep the herd off-limits from the tribe's daily needs, despite the obvious hunger among the Blackfeet.[19] In promoting a reservation stock industry, the OIA imagined the agency herd as an eventual source of cash income that would make the Blackfeet's annuity account self-supporting.[20] At the agency level, the cattle herd offered an opportunity to instruct the Blackfeet in a process of civilized production—a goal held in accordance with the OIA's broader policies of assimilation—that did not depend on raising crops across the reservation's cold and dry expanses. It offered a means to transform the Blackfeet from hunters to herders, from so-called predators to producers.

A lack of game in the early 1880s made the new cattle issue critical. During the summer and fall of 1882 Blackfoot hunters located only a few scattered herds of bison south of the Missouri River. Killing what they could, the hunters returned to their winter camps on their American and Canadian reservations. Short of meat, the Blackfoot survived the winter dependent on rations issued by their respective federal Indian agents and on any agency cattle they surreptitiously slaughtered. In the spring of 1883 they had once again left for their hunts, but they found almost no bison at all. Instead, they killed other game—elk, deer, antelope—along with a number of wandering cattle owned by white stock raisers.[21]

Conditions on the reservation grew increasingly grim over the course of 1883. In July OIA inspector Samuel Benedict reported: "No where in my journeying have I seen a country so destitute of wild game, you will ride for miles without seeing even a bird, all animal life seems to have either abandoned the country, or has been sacrificed to assuage the pangs of hungry men."[22] Over the next few months, the hunger only increased. Despite Benedict's pleas for the timely delivery of beef and flour rations owed to the Blackfeet for their recent land cession, freighters from Fort Benton, employed by former whiskey trader T. C. Power, failed to make the hundred-mile trip that summer and fall. By November the arrival of snow and ice had made the deliveries impossible, and the Blackfeet, living on the remote, windswept backbone of North America, entered the winter months desperately short of food. The Blackfeet who survived did so on an official daily ration of one-fifth of a pound of beef and one-third of a pound of flour, a winter diet of less than a thousand calories a day.[23]

The Blackfeet's pangs of starvation greatly interfered with the OIA's plans to assimilate the Blackfeet through the cattle herd. In desperation, the Blackfeet, along with agency employees, began slaughtering the agency herd for food. By 1883 only 132 cattle out of the original issue of 556 remained alive; 704 were killed during one year alone.[24]

In the midst of this destruction, Inspector C. H. Howard of the OIA visited the Badger Creek Agency and recommended the immediate construction of the slaughterhouse. Finished the following spring, the facility provided a means to more easily regulate the slaughter of agency cattle, and shortly after its completion, the unofficial killing of agency cattle grew increasingly criminalized. By the early 1890s, agency administrators had convicted so many Blackfeet of killing agency cattle—animals owned, of course, on behalf of the tribe's enrolled members—that the incidents greatly decreased.[25]

The convictions of Blackfeet accused of killing nonagency cattle also grew more serious. By the early 1880s Montana's growing cattle enterprises sought a spatial fix to the overaccumulation and devaluation of beef by literally moving their capital to greener pastures—the undergrazed grasses of the Blackfeet Reservation. On the eve of the 1883 starvation winter, I. G. Baker moved twelve thousand head of cattle onto the reservation's southern grasslands. Baker, a prominent Fort Benton merchant, was also responsible for fulfilling the agency's flour deliveries that winter, a task he failed to complete, exacerbating the Blackfeet's dire circumstances. Not surprisingly, the starving Blackfeet began killing his cattle, which were, after all, grazing illegally on their reservation. In response, an outraged Baker, along with other Bentonites who had lost cattle that winter, sent the Choteau County sheriff to the Blackfeet Agency, demanding the arrest of numerous Indians and Agent Young, who had been indicted by a grand jury. Young escaped arrest by handing over the Blackfeet suspects, who were taken off the reservation to Fort Benton and imprisoned, quite illegally, in the territorial penitentiary.[26]

The incarceration of these Blackfeet men amounted to the criminalization of traditional subsistence hunting. In the absence of other game, these hunters killed and ate Baker's range cattle, animals that wandered the reservation without Baker's supervision but that he still owned as private property. Even if the Indians were not actually killing cattle, their preferred mode of subsistence troubled the OIA and the regional cattle industry. After another catastrophic winter in 1886, for instance, when thousands of white-owned range cattle froze to death in northern Montana, hungry Blackfeet men left the reservation to search out the carrion. Salvaging this meat was a sensible proposition, but local cattlemen quickly devised a less sensible alternative. They hired skinners to take as many hides as possible, leaving the meat to rot. Making meat was to be a process of wage labor, not subsistence scavenging of the type conducted by wolves picking at carcasses.

Agent Young agreed. In the wake of the OIA's 1878 circular, which required all reservation Indians to perform labor in order to receive food rations, Young had protested to the commissioner of Indian Affairs that instituting this reform on the Blackfeet Reservation would prove impossible as long as Blackfeet men continued to hunt. He had long expressed his frustration with the Blackfeet men's seeming unwillingness to labor, "the men holding the notion that work is for women only and so long as hunting obtains subsistence it will take time, patience, and good example to introduce better ideas and practice."[27] The circumscription of subsistence hunting and scavenging forced the Blackfeet to work for wages or, at least, rations tickets. Incorporating the slaughterhouse as a male space that required male labors of distribution was an important method by which subsequent agents undermined the male hunting tradition. By establishing labor rather than predation as the essential input to making meat, OIA authorities could gender labor as male.

By the late 1880s the first slaughterhouse had not only confined killing and meat production but also grown into a reservation institution for male wage labor. The agency hired Blackfeet slaughterers and butchers—paid an annual salary of $500, mostly in rations tickets—to process about twenty cattle per week, an impressive number given the confines of the six-hundred-square-foot facility. These slaughterers worked under the supervision of the agency farmer, usually a local white man hired for the job. To guard against stock losses, the agency also hired a handful of Blackfeet men to work as full-time herders along with an Indian police force charged mainly with the task of arresting Blackfeet who continued to hunt their own beef. Remuneration for these jobs was less lucrative than that for the slaughterers (members of the tribal police force, for instance, received only extra rations tickets).[28]

Agency officials were at first quite pleased with Blackfeet workers at the slaughterhouse. The OIA approvingly reported that at Badger Creek, "beef is kept in better order than at any other agency." Four years later, another inspection revealed that the "butchering was more neatly and cleanly done than at any other agency in the service, the cattle being killed the day before issue and hung up over night to allow all animal heat to escape. During the summer they have butchered twice a week, and in consequence have had good healthful beef during the hot weather." The slaughterhouse, it seemed, was cleaning up the Blackfeet's barbarism and was leading them toward a civilized, healthy lifestyle.[29]

By 1892 attempts to transform the Blackfeet Reservation into a grand cattle ranch also seemed to be paying off. Even though Indians on the reservation still faced chronic food shortages, the agency sold its first cattle on the Chicago beef market.[30] Although these animals were shipped off for slaughter elsewhere, the agency slaughterhouse had allowed the herd to grow. By establishing the slaughterhouse as the only legitimate location for killing beef, the agency consolidated its control over Blackfeet subsistence. The circumscription of Blackfeet hunting accompanied the first slaughterhouse's transformation of Blackfeet meat production. The reservation's second slaughterhouse would reorient meat's distribution and consumption.

Clean Meat and the Second Slaughterhouse

In 1895 the Blackfeet Agency moved to Browning, Montana, and here the OIA built a new slaughterhouse, as well as a separate butcher shop. The move reflected a growing concern for modern and sanitary methods for slaughter. Moreover, by spatially differentiating the work of killing in the slaughterhouse from the work of preparing meat in the butcher shop, the agency helped cleanse meat from its violent origins, a prerogative held in accordance with the OIA's assimilative goal of unbarbarizing the Blackfeet. Since butchering meat had long been a labor performed by Blackfeet women, much to the chagrin of their Indian agents, the regulation of butchery within the confines of the butcher shop allowed OIA administrators to further supervise the Blackfeet's assimilation toward Anglo-American standards of gender and labor.

In contrast to the early optimism, by the early 1890s complaints had arisen regarding sanitation in the first slaughterhouse. Agents and inspectors commented that the old, hewn-log slaughterhouse was too small to accommodate the agency's needs and that it lacked adequate facilities to dispose of the various animal wastes that accumulated during periods of intensive use, amounting to around a thousand cattle per year.[31] Without pressurized hoses or concrete floors, removal of blood and organs from the log building presented a major challenge. Waged Blackfeet laborers, and sometimes prisoners in the agency jail, were faced with the task of maintaining a drainage ditch to carry away offal from the slaughterhouse.[32] Nevertheless, dried blood and rotten bits of flesh and guts supported a menagerie of insects and other creatures within the building's decaying timbered walls. One inspector remarked that the sheer amount of vermin sheltered in agency buildings would require a "sweeping fire to entirely exterminate." This wish was nearly granted in 1889, when a fire that erupted in the agency boarding school spread throughout the rest of the stockade but, ironically, spared the slaughterhouse.[33] By 1893 the OIA had insisted on its replacement, declaring that a "new slaughterhouse is needed where cattle for issue can be slaughtered by a scientific and civilized process."[34] Sanitized slaughter was becoming a clear concern.

However, concerns over health and sanitation were still bound with perceptions of the Blackfeet as vicious predators. Dating back to his first recommendations for an agency slaughterhouse, Inspector Howard, for instance, found traditional Blackfeet methods of slaughter, their "old, degrading, and disgusting practices," extremely disturbing, much more so than any other indigenous customs present on the reservation. Prior to the construction of the slaughterhouse, the Blackfeet continued to kill and process cattle—to make meat—using many of their former bison-hunting practices, revealing the agency's lack of progress in purging the Blackfeet of their seemingly predatory foodways. Rather than stunning and bleeding cattle, a method of slaughter standardized across most of Europe and North America by the late nineteenth century, the Blackfeet instead ran the agency cattle before killing them, in which case the animals would bleed out more thoroughly through their mouths following a fatal wound to the heart or lungs. And after killing the animal, the Blackfeet would rotate it on its back, peel away its skin and hide, and disembowel and butcher it on the ground—always outside. According to traditional practice, they would consume raw marrow from the leg bones while quartering the animal and also save its hooves, entrails, pooled blood, and head meat.[35]

The slaughterhouse was supposed to offer a major departure from these older practices, but by the 1890s it was clear that the fictions of humane treatment did not correspond to slaughterhouse realities. In 1893 a final inspector declared his outrage with the manner in which cattle were slaughtered at the agency. "At present, cattle are caught by casting a lasso around their necks," declared the inspector, "and in a choking condition drawn by a windlass into the door of the slaughter house, where they are pounded over the head until they fall to the delight and amusement of the Indians."[36] Not only confining killing but also further moderating the form that killing took became a major concern for the OIA's assimilation project. Even with the slaughterhouse, the boundaries between meat, labor, and killing were not well enough established.

Reconciling this ambivalence required a further separation of the labor of making meat from the nonlabor—the "delight and amusement"—of hunting and killing, and this was accomplished at the new agency by the construction of a butcher shop. After slaughtering and quartering, edible parts of the carcass were transferred to the butcher shop, where they were refashioned into meat suitable for distribution as rations. This physical removal and remaking provided an additional layer in the dissociation of meat from its origins in the flesh of the agency herd. The butcher shop also provided an improvement in terms of the aesthetic quality of the finished meat. Prior to its construction, agency employees chopped the beef with axes, using two logs as a table. This differed little from traditional Blackfeet methods of cutting meat, and disgusted OIA inspectors reported that "by the time it is ready for issue, it is in a horribly mangled condition."[37]

But this problem transcended a mere point of aesthetics. The butcher shop, like the slaughterhouse before it, provided another facility from which to regulate the distribution of subsistence to the Blackfeet. In this case, however, the agency sought to regulate not just the distribution of healthful meat but also the consumption of the "animal refuse" commonly eaten by the Blackfeet that OIA administrators deemed unhealthy. This was another problem that had accompanied the first slaughterhouse and one that persisted, especially while the Blackfeet remained hungry and sought to utilize as much of their beef cattle as possible.

Although the history of the reservation's early slaughterhouses has been all but ignored by scholars of the Blackfeet, William Farr's collection of reservation photographs includes two remarkable images of the facilities. In the first photograph, titled *Reaching for Entrails*, a group of Blackfeet are crowded around the Badger Creek slaughterhouse, peering through holes in the walls, one man reaching his arm through a gap in the timbers (see figure 2). Trying to pull something from the building's interior, the man's image documents the combination of resourcefulness and desperation that characterized reservation life in the 1880s. This kind of "reaching for entrails" is exactly what the OIA intended to stop with the construction of the second slaughterhouse. Framed walls would exclude the grasping hands of hungry Indians from the slaughterhouse's interior, where the by-products of meat making could be wholesomely eliminated. But as figure 3, *Agency Slaughterhouse, 1905*, indicates, the second slaughterhouse was probably more effective at exclusion than sanitation. A portrait taken within the Browning slaughterhouse of two men—one white and one Native—posed next to the skinned hindquarters of a steer lifted on a metal hanger, the image reveals only a mop and a bucket of water as a cleaning mechanism.[38]

For the OIA, clean meat was not necessarily antiseptic meat but meat from appropriate parts of the animal that had been prepared in acceptable ways by acceptable people. The cuisine created from what many whites considered animal garbage shocked administrators, despite the irony that

2. Reaching for entrails at the agency slaughterhouse. Courtesy of William Farr.

by the end of the nineteenth century, more and more Americans were eating meats like bologna, synthesized from the by-products of industrial slaughterhouses in Chicago and elsewhere. In all likelihood, the Blackfeet fed some beef entrails to their domesticated dogs, such as the ones surrounding the slaughterhouse in *Reaching for Entrails*. But even further turning the stomachs of colonial administrators was the fact that the Blackfeet women held a dominant role in the butchering and preparation of meat, particularly in the cooking of these more questionable meals, whether for dogs or for people. OIA inspectors were horrified to learn that on days of cattle slaughter, "women and children [were] allowed to hover around and paddle in the blood and garbage ad libitum." Once again, official concerns about Blackfeet modes of meat making combined the sanitation question of how to remove rubbish with the colonial question of how to properly manage Blackfeet civilization through their eating and labor practices.[39]

"Keeping the meat clean" entailed much more than properly bleeding, drying, and hanging animals; it also required setting boundaries on a social space that forcibly transferred the distribution of meat from the female to

3. *Agency Slaughterhouse, 1905*. Courtesy of the Sherburne Collection, Mansfield Library, University of Montana.

the male domain. As plans for the first slaughterhouse were drafted in 1883, one inspector suggested that "no women or children should be permitted about the place." Instead, work at the facility was to be a male enterprise: "A delegation of men from each village might be selected to receive such of the refuse as can, in any possible way, be converted into food. The police could be used to see to an equitable division of these 'spoils.' The meat should be kept clean, it is now in a disgracefully filthy condition." Although the exclusion of women was not fully possible following the construction of the first slaughterhouse, the physical separations inherent to the second slaughterhouse and its accompanying butcher shop assisted the OIA's long-standing efforts to transform female Blackfeet labor.

Agency administrators also worked to institutionalize this transformation in a series of attempts to refashion Blackfeet women from butchers into bakers. The fundamental element at stake in this transformation was the Blackfeet diet, which had been long dominated by animal proteins and which had no analogue to bread. Instructing the Blackfeet on how to eat and bake bread became an obsession at the agency boarding school. In 1893 the Blackfeet agent successfully lobbied the OIA to construct an expanded bakery at the new Browning Agency. One of his main concerns was the ability of the agency bakery to produce enough good-quality bread.[40] The OIA also attempted to supplement smaller beef rations with weekly issues of flour. Without access to other essential baking ingredients such as yeast and baking powder, most Blackfeet simply consumed the flour plain, likely in some kind of porridge—a circumstance that converted few Blackfeet to the white man's favorite food.[41]

Although the second agency slaughterhouse first emerged from concerns over sanitation, its larger evolution grew from OIA urges to keep meat clean by disavowing its predaceous origins and by limiting female control over its preparation and consumption. By building a separate butcher shop, the agency added an additional layer of mystification between meat and its relation to animal flesh. The separation of slaughter and butchery also enabled more effective supervision over the circulation of animal by-products that represented an important component of the Blackfeet diet. By restructuring Blackfeet foodways to suit the objectives of federal Indian assimilation, the slaughterhouse provided a key institution for colonial policy.

Conclusion

Assimilation policy on the Blackfeet Reservation was foremost oriented toward subordinating Blackfeet land and labor into northern Montana's capitalist livestock industry. This required not only policing the reservation but also changing what the Blackfeet ate, along with how they made that food. Transformations in Blackfoot ways of making meat played a pivotal role in these policies of settlement and assimilation during the late nineteenth and early twentieth centuries. During these years, the lives of many Blackfoot people transitioned from hunting to wage earning, and making meat changed from a direct procedure of killing animals and eating them into an indirect process focused on private property and market exchange, dominated by the region's white cattle growers. Subordinating Blackfoot land and labor was a critical step in turning the northwestern plains into productive cattle country, and taking command of Blackfoot ways of making meat offered the key to achieving this goal.

Establishing and negotiating the contradictions of capitalist enterprise was yet another goal of assimilation policy on the Blackfeet Reservation. At the heart of these contradictions was a tension over valuations of labor as communal or individual. Concealing the contributions of human and nonhuman others within the "self-making" ethos of individual capitalism bolstered colonialism's discourse of production. By casting the work of raising cattle as productive and the labors of hunting bison and other game as predatory and nonproductive, the Office of Indian Affairs sought to subordinate Blackfeet practices of making meat to conform with the growth of a regional capitalism based on beef. Subsuming the nonhuman labors of livestock under the labor of individual stock growers required a disavowal of animal labor, a process that worked out on the Blackfeet Reservation through OIA attempts to dismantle Blackfeet practices of nonhuman personhood and human-animal kinship. While this chapter examined OIA efforts to deanimate Blackfeet ways of making meat, the next chapter demonstrates how the Blackfeet accommodated their own understandings of labor and human-nonhuman relationships to OIA prerogatives. Despite colonial efforts to displace these concepts, animistic relationships still pervaded capitalism's expansion across the Blackfeet Reservation.

In both Montana and Alberta, the circumscription of reservation hunting practices quickly gave way to ration-ticketing systems that encouraged "productive" labor—not subsistence labor, but labor that benefited the federal agency, labor that could be exchanged indirectly for life's necessities. The history of the Blackfeet Reservation's agency slaughterhouse provides a key example of these transformations. By using the slaughterhouse to regulate access to meat, agents of the U.S. Office of Indian Affairs believed they could transform the Blackfeet from hunters to herders, from barbaric predators preying on the plains' ownerless stocks of animal capital to civilized producers subject to Anglo-American standards of labor, property, and land tenure.

Meat's relationships with colonization reveal connections between the dispossessions of Blackfeet land and labor. As a physical and cultural link between land, bodies, and work, meat adjudicated understandings of human and animal difference, of race, of gender, and of civilization and barbarism. Regulating access to meat provided the OIA an important means to transform the Blackfeet into colonial subjects.

OIA attempts to subordinate and reform Blackfeet labor focused on transforming Blackfeet foodways. By 1895 the new slaughterhouse and butcher shop at the Browning Agency had instituted profound changes in the means by which the Blackfeet acquired their subsistence. Trading work for prepared meat dissociated from its animal origin had reoriented the process of making meat from a collaborative project of hunting and butchering bison to an individuated procedure of exchanging labor for beef. By putting hunting under siege and by forcing a separation of "productive" labor from the predatory nonlabor of hunting, the slaughterhouse provided the means by which the OIA could enforce its colonial dictum to either "work or starve."

The agency slaughterhouses also fit into larger patterns of the late nineteenth-century livestock industry. Like northern Mexico, the central plains, and other dry, grassy interiors of North America, cattle raised on the northern Rocky Mountain front mostly fed urban consumers. Until the 1890s the Blackfeet Reservation was an exception to this rule. A long-standing hunting tradition, along with the demands of hunger, brought the Blackfeet to kill their cattle to satisfy their immediate needs rather than the needs of a distant marketplace. The agency slaughterhouse proved instrumental in disciplining this aberration. By restructuring Blackfeet foodways, the agency's slaughterhouses opened the way for the reservation's integration as a beef production outpost within a global food system that channeled profits to large owners of animal capital.

Under the pressures of federal assimilation policy, food not only fueled labor but also helped determine what counted as labor. The transformation of the Blackfeet into colonial subjects relied in part on OIA authority over beef, which could not be hunted, only rationed out or purchased. The OIA's assimilation policies sought to incorporate the Blackfeet as unequal participants in the region's broader cattle economy. These policies hinged on understandings of predation and production that not only disciplined Blackfeet deviations from Anglo-American standards of meat making but also undergirded colonial notions of labor and land tenure. On the Blackfeet Reservation, beef compelled a new way of life alongside a new basis of subsistence.

4

The Place That Feeds You

The winter of 1919 blasted an exodus of settlers from the Northern Rockies frontcountry. A crippling drought had broken that November, but it was immediately followed by a series of eastward-rolling storms that cemented the range with a glaze of ice through April. Stock growers spent the winter miserably awaiting the deaths of their starving and freezing cattle. Those who could afford hay paid up to five times its usual price. Others simply loaded their underweight steers on railcars to Chicago, selling them at a loss. As the spring thaw commenced, the saturated beef market crashed when stock growers across the northern plains sold out their herds to cover enormous debts. Meanwhile, postwar price depressions in wheat and barley bankrupted the region's dryland farms, already suffering from the ecological repercussions of bonanza agriculture. At the same moment that Europe's granaries recovered from the shock of total war, the high plains' fragile soils shuddered under aggressive bumper crops that had reduced entire fields to "a solid mat of Russian thistle," as Wallace Stegner remembered the fate of his own childhood homestead, just a few miles above the Montana border. Along with thousands of other settlers, Stegner's family fled the landscape they had desiccated; by the winter of 1920, few remained.[1]

These combined misfortunes of climate, commerce, and industrial agriculture ravaged the Blackfeet as well, but, unlike their Anglo-American neighbors, the Blackfeet could not pick up and move their capital to greener pastures. Incorporated into the free market of twentieth-century agriculture under legal conditions different from those of white homesteaders and stock growers, most Blackfeet could not sell their reservation land allotments until 1918, and only then with the approval of Office of Indian Affairs (OIA) administrators. These federal agents determined which individual Blackfeet were "competent" to own fee title to the hundreds of acres of land that the federal government had held for them in trust since the reservation's allotment in 1907. With few other options, most Blackfeet remained on the reservation and continued living day to day on OIA food rations and on wildlife and cattle that they hunted.

The political ecology of the reservation changed dramatically in 1921, however, with the establishment of the Piegan Farming and Livestock Association (PFLA), a collection of twenty-nine Blackfeet-led farming districts across the reservation organized around the cooperative use of individual land allotments for the common goal of reservation food security. By 1923 the Blackfeet were raising enough chickens, pigs, and cattle and growing enough wheat, potatoes, and other vegetables through the PFLA to satisfy their own annual needs, as well as some market demand off the reservation. During a decade when industrial agriculture in Montana, Alberta, and Saskatchewan faced economic crises of unprecedented magnitude, the Blackfeet had organized an alternative agricultural community that sustained its members through an emphasis on the local production, distribution, and consumption of mixed food crops.

This local understanding of sustenance had long been an essential component of Blackfeet conceptions of home and territory. The Blackfoot word for territory, *áuasini*, translates into English as a combination of food and place, something like "the place that feeds you." The word expressed broader Blackfoot knowledge of human-nonhuman kinship and relational selfhood, an acknowledgment that one does not autonomously feed oneself but necessarily relies on a wider set of social, animal, and environmental relations to acquire sustenance. It drew on Blackfoot understandings of selfhood that were unlike the production-focused individualism residing at the core of OIA assimilation policies during the late nineteenth and early twentieth centuries. Rather than forging an autonomous selfhood through a series of dualisms that abstracted persons from their environments, as colonial capitalism sought to effect, Blackfoot knowledge combined embedded selfhood within concepts of place. So too did production and consumption coexist as inseparable processes. Value circulated and recirculated throughout a world of kinship obligations held between human and nonhuman persons. These Blackfeet understandings revealed predation as a cultural concept specific to Anglo-American colonialism, an ontological structure that worked to determine the legitimacy of the individuated labors of abstracted selves. In contrast, according to Blackfeet indigenous knowledge, predation and production were mutual concepts.[2]

The PFLA emerged as an expression of two decades' worth of organized Blackfeet resistance to OIA cattle-leasing and allotment policy. The accommodations to colonialism worked out by the Blackfeet had hinged on mobilizing Blackfeet understandings of relational selfhood to work within the exigencies of modern colonial capitalism. In spite of criticism that depicted the PFLA, alternately, as both "old-fashioned" and "Bolshevistic," the project earned praise for establishing a level of food security unheard of in Blackfeet country since the 1870s. Indeed, thanks to the PFLA's success in incubating Blackfeet notions of place within the exigencies of allotment and federal assimilation policy, by 1924 its Blackfeet members claimed that they had become "the most progressive tribe of Indians in the United States."[3]

With the PFLA, the Blackfeet were not simply "reviving tradition" but articulating a new vision of relational selfhood during the Progressive Era that drew on indigenous knowledge to understand individual livelihood in the modern world as part of a cooperative enterprise that spanned the labor of human and nonhuman persons. Since the 1880s the OIA had worked to divest the Blackfeet of their tribal obligations and assimilate them as individuated selves through institutions such as the slaughterhouse. Federal authorities tried to conduct many of these assimilation efforts in good faith throughout the end of the nineteenth century, as Frederick Hoxie and other historians of U.S. Indian policy have demonstrated.[4] But this push ironically coincided with a historic moment in which the very model of autonomous selfhood capitulated under the tide of corporate-industrial capitalism, a transformation studied by historians such as James Livingston and Jackson Lears.⁵ Drawing on their own indigenous knowledge of kinship and selfhood and on the lessons learned from their experience with modern agricultural, the Blackfeet were in a suitable position to address the challenges that drove so many of their white neighbors from the land. They realized, as well, that the cooperative use of their privatized land resources provided a meaningful way to develop economic independence and community sovereignty in an era when mass-market forces reshaped social relationships throughout North America.

The previous chapter examined the role of the reservation's slaughterhouse as an institution through which the OIA attempted to subordinate Blackfeet labor and circumscribe hunting, procedures that hinged on Anglo-American visions of purging the Blackfeet of their predatory foodways. This chapter traces the evolution of Blackfeet notions of place and kinship in structuring Blackfeet resistance to OIA colonialism in the early twentieth century. Intratribal debates over OIA grazing leases and reservation cattle trespass led many Blackfeet increasingly to question Anglo-American standards of selfhood, labor, and value during the first decade of the 1900s. One of the leaders who emerged from these debates was Robert Hamilton, whose tireless campaign against the Allotment Act's surplus land sale provision succeeded with its repeal in 1919. Hamilton's focus on land as the basis for Blackfeet economic independence drew on a wellspring of Blackfeet notions of áuasini that informed the PFLA during the 1920s. As the capitalist dissociations of industrial agriculture collapsed around the postwar northern plains, the PFLA offered the Blackfeet an alternative means to reconfigure their world.

This history of the PFLA's successes, however, has been mostly overlooked in the historiographies of both northern plains agriculture and the Blackfeet. The PFLA's prioritization of self-reliance over commercial sale marks the organization as an imperfect fit in the industrial narrative of northern plains agriculture in the early twentieth century, a history that has been well documented by historians of environment and technology such as Deborah Fitzgerald, Mark Fiege, and Rod Bantjes.⁶ Yet despite its crucial departures from industrial, market-based agriculture, the PFLA bore certain similarities to the organized factory farms, weed control districts, and farmers' cooperatives that these and other historians have studied. Like other Progressive Era agricultural associations, the PFLA responded to Montana's agricultural challenges in the 1920s by encouraging social cooperation. Far from developing in an isolated colonial vacuum, the PFLA emerged as part of a larger regional, national, and transnational trend toward the social reorganization of farming in the decade following the First World War.

Several generations of American Indian historians have now studied the PFLA, approaching it mainly as a bureaucratic anomaly of early twentieth-century Blackfeet history. John Ewers, Paul Rosier, and Tom Wessel, working from mostly OIA sources, have ascribed the PFLA's successful emergence primarily to the good-hearted efforts of Fred Campbell, the reservation's superintendent throughout most of the 1920s.[7] They are right to point out that Campbell's earnest administration represented a radical departure from the corrupt and inefficient tenure of previous Blackfeet agents, particularly from 1916 to 1920, when no fewer than seven agents cycled through Browning, an era that Rosier calls the "revolving door period."[8] But the PFLA, born from Campbell's ambitious Five-Year Industrial Program, succeeded in spite of the OIA's economic motivations for the reservation. It succeeded because it offered the Blackfeet opportunities to refashion their indigenous understandings of production and relational selfhood into a viable and modern social alternative to the forms of commodity agriculture that had bankrupted the frontcountry over the previous decade.

This oversight is also due to the fact that ever since the industrialization of North American agriculture in the early twentieth century, agricultural historians have been held in thrall to the logic of predation that has structured their subsequent understandings of what constitutes modern agriculture. As discussed in chapter 2, this logic served to propel the cultural ascendance of industrial stock raising in places like northern Montana and southern Alberta both by fixing visions of productive labor to the humanized figure of the cattleman and by animalizing the so-called predatory labors of wolves, Indians, and others. As chapter 3 explained, this logic also worked to circumscribe Blackfeet foodways on the reservation throughout the OIA's assimilation campaigns of the 1880s and 1890s, first by identifying Blackfeet ways of making meat as predatory and then by dissociating meat from its animal origins—via the slaughterhouse—in attempts to make Blackfeet wage labor the basis for legitimate production. A similar process accompanied the industrialization of agriculture across the northern plains, particularly wheat farming. As the Farm Bureau, the Nonpartisan League, the State Extension Service, and other associational entities worked to bind industrial wheat production to the social core of national agriculture in the 1910s and 1920s, they did so in the various contexts of designating competing interests as both metaphorical and literal predators—in some cases banks and railroads, in other cases immigrants and squatters. During these years, the disavowal of one's own predatory inclinations proved key in establishing legitimate claims as a producer. Once cordoned off as an industry of production aligned primarily with wheat monoculture, the category of northern plains agriculture remained, at best, mostly off limits from mixed-farming projects like the PFLA. And at worst, according to these conceptions, the PFLA represented an unacceptable bricolage, a Bolshevistic anathema to the prevailing industrial order.

Cattle and Labor

By the early 1900s the growth of the reservation cattle industry had begun to replace horse ownership as one of the main sources of class division on the Blackfeet Reservation. For the next thirty years, battles between the reservation's procattle and anticattle factions dominated debates over the future promises of Blackfeet national sovereignty and livelihood. Different understandings of predation, labor, and the origins of value lay at the center of these controversies. Different visions of land, cattle, and labor animated discussions of Blackfeet nationhood. Blackfeet cattlemen, with key support from the OIA, imagined the Blackfeet Reservation as a vast grazing reserve where individual Indian stock growers would base their economic independence on the market sale of beef cattle. Other Blackfeet objected to the privatization of the reservation's animal capital and emphasized tribal obligations to community well-being over the accumulation of individual fortunes in beef. Blackfeet understandings of labor, kinship, human-animal personhood, and relational selfhood stood at the heart of these objections, while an ethos of pioneering individualism pervaded the reservation's cadre of cattlemen and their allies at agency headquarters.

The expansion of the cattle industry accentuated the class divisions that emerged on the reservation during the late 1800s. In 1902 an OIA survey revealed that only 572 Blackfeet out of a population of over 2,000 owned any cattle, while 47 owned over 100 head each.[9] Intratribal divisions over cattle racked Blackfeet politics during the early twentieth century. With cattle ownership increasingly concentrated among a core elite of reservation stock growers near the agency headquarters in Browning, the OIA forged key alliances with Blackfeet ranchers in their efforts to privatize reservation landownership through the allotment process. However, opposition to allotment mounted primarily by non-cattle-owning Blackfeet succeeded in halting the sale of so-called surplus unallotted land under the Blackfeet's original allotment act. Opposition to allotment also provided the political mobilization necessary for the development of the Piegan Farming and Livestock Association, which advocated mixed subsistence agriculture as an alternative to a reservation export economy oriented around the sale of marketable beef cattle.

For the OIA, cattle remained the key for assimilating Blackfeet livelihood to Anglo-American standards of production and individualism and for subordinating Blackfeet land and labor within Montana's industrial livestock economy. Agent George Steell exemplified the OIA's faith in the revolutionary power of cattle in remaking the Blackfeet: "The possession of cattle and the proper care and protection of this stock will make the man out of the Indian.... [T]he influence of the Chiefs and Medicine Men will disappear."[10] Individual ownership of cattle and the husbanding of those animals would purge the Blackfeet of their predatory inclinations to hunt. Following in the path of the agents who preceded him in the 1870s and the 1880s, Steell was further convinced that industrious cattle raising was the key to assimilate the Blackfeet into the American capitalist body politic.

But even the successful establishment of a reservation cattle regime that benefited the Blackfeet rather than the reservation's non-Indian neighbors faced substantial obstacles. Foremost among the problems was livestock trespass. For decades, white cattlemen had trespassed sheep and cattle on the reservation, sometimes with hundreds or even thousands of head at a time. Since the reservation was largely unfenced, neighboring herders had little to fear from charges of trespass. If caught and confronted by the reservation's tribal police or its few line riders, the herders could usually get off by claiming their stock had strayed onto the reservation by accident, at which point the Blackfeet were obliged to assist in rounding up and herding the trespassing stock back off the reservation. The lack of a suitable boundary fence made it difficult, if not impossible, to prove if a stock raiser had herded animals onto the reservation unless he was caught in the act.

Even when trespassing cattlemen were prosecuted and OIA agents levied fines against the offenders, the payments were rarely collected. The political will necessary to punish and prevent cattle trespass did not exist in Browning any more than it did in neighboring county jurisdictions. For instance, in 1892 Steell wrote a stern warning to the trespassing foreman of the Flowerree Cattle Company, notifying him that "I will be compelled to take steps towards collecting for the grazing of these cattle upon Indian lands. In the future you must see that your cattle do not infringe upon the rights of these Indians."[11] The fines were never collected, and Steell was himself convicted of cattle trespass a year later, when other employees at the agency reported that he brought his own private herd to graze in the agency enclosure. Cattle trespass was widespread, difficult to control, and present even within the highest levels of reservation's bureaucracy.[12]

Accepting the realities of reservation cattle trespass, the OIA began a policy of leasing portions of the Blackfeet Reservation to non-Blackfeet grazers in 1904, a proposition that at least garnered some income from the Montana cattlemen who continually turned stock loose on the reservation. The decision to lease the reservation also led to the construction of a much-needed boundary fence, which was completed in time for the first lessees toward the end of 1904. Not surprisingly, the reservation roundup conducted that fall discovered a large number of off-reservation cattle of various ages that had been grazing on the reservation for quite some time, all of them bearing the brands of prominent local stock growers, many of whom had signed on as the reservation's largest lessees. A report filed with an auditor from the Treasury Department in 1907, for instance, showed Dan Flowerree with six thousand head of cattle on the reservation, William Wallace Jr. with forty-five hundred, and the Sun River Stock and Land Company with fifteen hundred cattle and five hundred horses. These animals greatly outnumbered the Blackfeet's small tribal herd, as well as the private herds of the larger Blackfeet stock growers. Less than a dozen Blackfeet ran herds larger than one hundred head, and none owned more than a thousand animals.[13]

Although this leasing program generated revenue for the agency, it also created a number of problems. First, Blackfeet owners of private herds protested the presence of outside cattle grazing on reservation grass. They correctly alleged that the new leasing policy was a scheme to abandon the OIA's responsibility of managing tribal lands for the benefit of Blackfeet-owned cattle in order to raise revenue for the agency at the expense of Blackfeet stock growers. They also protested the measure in light of the fact that they had been paying grazing taxes for their own private cattle to the agency since the 1890s. The leasing policy with outsiders left Blackfeet grazers without any real advantage over the white cattlemen who were stocking the Blackfeet Reservation with thousands of cattle.

Somewhere, the Blackfeet asserted, agency policy needed to change, and the largest Blackfeet cattlemen fought vociferously to end the grazing tax on private Blackfeet cattle, winning OIA agents to their side. To that end, they protested the subclassification of reservation cattle that left their herds subject to taxation. Cattle issued to Blackfeet individuals by the OIA under treaty were branded ID for Indian Department and were not subject to the tax, while "straight" cattle, those bought off-reservation by enrolled Blackfeet and given their individual brands, were taxed. Beginning in 1904, all three Blackfeet agents during this era, J. Z. Dare, Clarence Churchill, and Arthur McFatridge, simply refused to collect grazing taxes, which transformed into a hardened dispute with their Indian Office superiors. The spirit of the tax, claimed Blackfeet cattlemen and their agency allies, seemed to contravene the OIA's mission to assimilate the Blackfeet as capitalist stock raisers. As McFatridge explained, "To me it appears to favor those who have obtained cattle through no exertion of their own, nor expenditure of their personal funds, and puts a penalty on those who by industry and frugality have accumulated cattle which are not branded ID." But the OIA rejected proposals to end both the straight tax and the outside leases on the basis that both generated much-needed revenue to pay for the agency's routine operations.[14] OIA headquarters hewed to policy decisions that would make agency administration self-funding at the expense of aiding "industrious" Blackfeet stock growers, an outlook that must have infuriated Blackfeet cattlemen, as well as agency employees in Browning.

A second problem with the leasing program was how easily local stock growers could manipulate it. These cattlemen cooperated to bid grazing contracts with the agency down to the lowest amount possible—to the point where it was cheaper for white lessees to graze cattle on the reservation than for Blackfeet ranchers to graze their straight cattle. By 1908 grazing leases on the Blackfeet Reservation were significantly cheaper than on the Crow Reservation, where a similar program had been put in place. That year the Blackfeet Reservation's highest bids came in at $1.30 per head, a full 20 cents less than the grazing tax paid by Blackfeet ranchers on straight cattle. Agent Churchill seemed aware of this problem but was unable to do much about it, short of refusing to lease to anyone. "It appears to me that this offer is a concocted scheme on the part of all former permittees," he wrote to the commissioner of Indian Affairs, "procuring the maximum grazing for the minimum cost."[15] The OIA offered no effective response. Rather than setting a standard price schedule for grazing leases, the OIA continued its policy of open-market bidding for Montana cattlemen while subjecting Blackfeet stock growers to a static grazing tax on their own reservation.

Not surprisingly, lessees under the OIA's program also consistently cheated their leases by stocking the range with greater numbers of cattle than they were permitted. In 1906 Churchill reported his suspicion that there were around 10 to 15 percent more cattle on the reservation than officially reported. At that rate, lessees would have owed around an additional $1,500 a year, enough to pay the salaries of four additional agency farmers or livestock supervisors. This trespass was difficult to confirm, however, because the agency let its lessees handle the bulk of the roundup labor. The lessees were responsible for reporting the number of cattle they grazed on the reservation, since agency personnel alone could not tackle the enormous task of rounding up all the reservation's stock. With around twenty thousand head officially grazing on the reservation by 1907, the great majority were owned by off-reservation lessees.[16]

The geographic placement of these leases also posed a problem that would erupt in political divisions between procattle and anticattle Blackfeet.[17] The economic benefits of the leases were largely captured by the agency administration, while its environmental externalities fell on local groups of non-cattle-owning Blackfeet. Since the leases were organized by grazing districts, everyone on the reservation supposedly benefited from the revenue generated by the leases, but the ill effects of the lessees' huge cattle herds were confined to local populations. Generally, the communities hardest hit by these circumstances were poorer, cattleless Blackfeet living in the southern and eastern portions of the reservation rather than the relatively prosperous Blackfeet ranchers clustered in and around Browning. Blackfeet living on a denuded rangeland surrounded by thousands of white-owned livestock learned firsthand the aggravations of living near range cattle. The process of making beef on a grazing lease required the cosmic efforts of the sun and water, the metabolic efforts of grass and the cattle themselves, and the willingness of Blackfeet to suffer the social and ecological consequences of range degradation. The cattle leases left the Blackfeet with wrecked streambanks and piles of manure, the profits from the sale of beef raised on the reservation went to off-reservation cattlemen, and money from grazing leases filtered into the coffers of the Indian agent. By 1909 residents in the hardest-hit districts complained bitterly about these injustices to agent Churchill, who passed their sentiments on to the commissioner of Indian Affairs. The OIA nevertheless decided to continue the leasing program at least until the planned completion of the reservation's allotment in 1912.[18]

In the midst of these debates over trespass and leasing cases, the merciless winter of 1919 was foreshadowed by another brutal winter in 1906. To OIA personnel invested in the cattle industry, the winter seemed to turn back the clock. Many starving Blackfeet began scavenging meat from frozen beef carcasses. The next spring, agent Dare reported, "The loss of cattle during the past winter to both Indians and permittees, has been very heavy and many Indians have been subsisting from cattle that have died from the effects of the cold, and from the sale of hides taken from the carcasses of said animals."[19] That year's winter almost completely wiped out a large issue of ID beef cattle made to individual Blackfeet the previous summer. Purchased from the Texas panhandle, where a large glut of poorly bred beeves had driven down prices, the OIA shipped the cattle over a thousand miles to Browning, just in time for one of the worst winters in twenty years. Most of the unseasoned Texas cattle died. Three years later, Churchill expressed his doubts that "there are [even] five hundred head of the 2000 cows and calves issued now living. One mixed blood Indian who has been more successful in stock raising than the average tells me that some of this southern stock actually froze to death while standing in sheds knee deep in hay at his ranch." The great die-off temporarily crippled the reservation cattle industry, as even the largest stock owners on the reservation contemplated insolvency. But more importantly, the majority of Blackfeet cattle owners, most of whom owned fewer than half a dozen animals, were immediately devastated by the loss of their small herds. With few other sources of livelihood for the coming year, these people would end up back on the rations rolls or else stalking others' cattle, as OIA administrators anticipated.[20]

Within the next few months, Churchill was fired, a result both of his inability to prevent cattle theft and of a malicious petition signed by various white residents of Browning who claimed that Churchill tried to seduce the wife of the agency physician.[21] Early in 1910 he was replaced by Arthur McFatridge, a former boarding school superintendent on the Rosebud Reservation who promised an end to cattle "depredations" and the transformation of the Blackfeet Reservation into a productive landscape.

This was an empty promise both because the OIA did not follow through on its plans and because the Blackfeet resisted the OIA's efforts. As portly and pompous as his name might suggest, McFatridge sought to rule the Blackfeet Reservation as a vast cattle kingdom, a project he shared with his predecessors dating back to the 1870s. In the wake of the Great Northern Railroad, homesteaders, and the cruel winter of 1906, the Blackfeet

Reservation was all that remained of the open range in northern Montana, a land long celebrated for its unfenced grass. By 1910 neighboring white ranchers were paying $1.50 per head of cattle to agency personnel for the privilege of grazing their animals on the Blackfeet's tribally owned grasslands. Still living by their industry's obsolescent rhythms, they turned their stock loose to fatten and propagate across the boundless plain. But under the modified Allotment Act of March 1, 1907, this world would gradually come to an end. The reservation's allotment into private property had begun and would be mostly complete by 1913, leaving only a six-hundred-thousand-acre unallotted corner to serve as open rangeland.

Fearing another string of depredations like the ones of the preslaughterhouse days, the reservation's large stock growers and lessees agitated for greater enforcement of antirustling laws. Whether the incidence of cattle theft actually increased from 1907 onward or not, most reservation stock growers and off-reservation lessees seemed to think it had. In 1909 Churchill had reported that "cattle and horse thieves have been working on the reservation to an alarming extent the past year," even rustling off a herd of work horses used by the Bureau of Reclamation.[22] Some of the worst victims of theft were small Blackfeet stock owners with ID cattle. The unsophisticated ID brand was easily worked over by experienced cattle hands, probably some of whom already had practice doing so on other northern plains reservations like the Crow, Pine Ridge, and Rosebud.[23] In 1909 Churchill was reluctant to report that the theft of cattle had reached a fever pitch.[24] Unable to assuage the stock growers' concerns over theft, and unable to improve conditions for the increasing numbers of starving enrollees, Churchill resigned his position later that winter.

His successor, the notorious Arthur McFatridge, hastily formed the Blackfeet Stock Protective Association (BSPA) to address the problem of cattle theft. Seeking to run the Blackfeet Reservation as his own personal cattle kingdom, McFatridge envisioned himself as a twentieth-century successor to Granville Stuart, the wealthy rancher who cemented Montana's vigilante tradition by orchestrating the murders of alleged cattle thieves in the 1880s.[25] Harking back to the old-time vigilante groups that had terrorized Montana in the years after the Civil War, McFatridge organized the BSPA and charged it with protecting the reservation's animal property. With a contempt for legal process unusual by federal standards but typical by Montana standards, McFatridge deputized the BSPA's members, granting them the authority to summarily arrest and jail any individual caught stealing or killing cattle on the reservation. Forty-three men paid five dollars each for this power. McFatridge himself anxiously wrote his superiors in Washington DC, begging for the approval to grant this authority. Despite its auspicious start, the BSPA would not amount to much, other than a bullying organization for the reservation's budding cattlemen.[26]

Like Montana's wolf bounty laws, the BSPA's short-lived vigilante uprising against cattle rustlers reflected stock growers' desires to regulate the legitimacy of the labors of hunting and husbanding more than its actual effectiveness in eliminating fraud and theft. As such, the BSPA was a means to harness representations of predator and producer to justify the particular flow of value from cattle to their owners as much as it was intended to make sure that this flow actually did occur. This emphasis on regulating forms of labor rather than measuring labor in simple terms of subsistence and provision also mirrored the Office of Indian Affairs' wider approach to labor and assimilation policy on the Blackfeet Reservation. The OIA hoped to transform the Blackfeet from "idle tramps," as Senator Henry Dawes had described them in 1885, into sturdy, individual cattle growers, albeit not ones independent but rather dependent on markets controlled by a regional white elite. This was the essence of colonialism—the ruling class's disavowal of its own predatory inclinations through the adjudication of legitimate labor and nonlabor.

Despite these substantial challenges, from the 1890s through 1912, when the reservation's allotment was completed, Blackfeet stock growers amassed modest fortunes raising cattle on the reservation, building their herds through a series of federal decisions that favored their enterprise, through the purchase of other Indians' tribal issues of stock cattle, and through partnerships with white cattlemen living off the reservation. However, the majority of Blackfeet continued to struggle in the market economy. Mostly without cattle, without opportunities for employment, and lacking access to traditional sources of subsistence, these Indians lived without cash and remained on the rations rolls.

The Blackfeet sought their own modern solutions to these problems. At the heart of their considerations was a tension over valuations of labor as communal or individual. Concealing the contributions of human and nonhuman others to "self-making" of individual capitalists offered the main advantage of colonialism's discourse of production. But in seeking to disenchant and deanimate the origins of value in their representations of labor, capitalist stock growers nevertheless relied on animal labor and were forced to recognize this dependence in their day-to-day operations. The reservation's livestock regime existed in tension with Blackfeet knowledge of áuatsini, and controversies over allotment would further expose these divisions between the OIA's goals of cattle-based individualism.

Allotment and Land Sales

Cattle ranching and allotment politics made for strange bedfellows on the Blackfeet Reservation. Left completely unallotted until 1907, a full twenty years after the passage of the Dawes Act, the reservation had become a unique open range precisely because of its unallotted lands. As homesteaders carved up the Montana plains and plowed under its cattle trails, many local white ranchers viewed the reservation as the final refuge of open-range stock raising. OIA personnel were well aware of this fact, Clarence Churchill explaining that "stockmen who are acquainted with all the grazing sections of the State, admit that there is no better grazing in any part of Montana than the Blackfeet Reservation."[27] In managing the reservation's grasslands, the OIA was tasked with balancing the needs of the Blackfeet's tribal herd along with the needs of cattle herds owned by a few prosperous enrolled Blackfeet and, after 1904, the desires of local whites who began leasing significant portions of the reservation's acreage. This was a challenging job that the OIA was ill suited to undertake. The allotment process aggravated its difficulties.

As Blackfeet foodways shifted during the early reservation era, the Blackfeet adapted their visions of land and sovereignty to reflect changed circumstances in food production and distribution. John C. Ewers, the well-respected mid-twentieth-century Blackfeet historian, remarked that Blackfeet survival during this era—the 1870s through the 1920s—relied on a process of "trading land for food." According to Ewers, the major land cessions of this period, which reduced the reservation from around fifty thousand square miles to three thousand square miles—a difference approximately the size of Pennsylvania—were accepted by the Blackfeet because their vast eastern ranges no longer supported bison. And after the adoption of reservation stock raising in the 1890s, Ewers argued that the reservation's overhunted mountainous western fringe no longer mattered as a serious source of food production, so the Blackfeet were willing to trade it—it is now part of Glacier National Park—in exchange for future OIA cattle and rations issues. Blackfeet wealth in land, so goes Ewers's narrative, provided hard currency to weather a brutal modern American colonization that repeatedly forced the Blackfeet to the brink of starvation.[28]

Although more recent historians of Blackfeet politics such as Paul Rosier have been careful to point out that the difference between "full blood" and "mixed blood" designates cultural rather than biological distinctions, in the case of allotment, it may be useful to organize Blackfeet political history around supporters and opponents of the livestock industry who developed differing political identities based on their different understandings of place, livelihood, and selfhood—concepts tied directly to notions of predation and production. Following Ewers, historians have complicated Ewers's narrative, but mainly by recourse to race. In doing so, they have placed too great an emphasis on blood quantum as a denominator for Blackfeet land politics in the early twentieth century, situating controversies over allotment on a continuum between white and red. Although the majority of Blackfeet stock growers were "mixed bloods," it is misleading to categorize stock growers on that basis, since a handful of "full bloods" numbered with the reservation's largest cattlemen, including Wolf Tail, who also sat as a judge on the agency's Court of Indian Offenses.[29] Likewise, many "mixed bloods" resisted the OIA's management of their reservation as an industrial cattle outpost and sided with the reservation's anticattle faction, including Robert J. Hamilton, son of a Pikuni woman and an Irish whiskey trader and the ostensible leader of the "full bloods" through the 1910s.[30]

By 1912, when Blackfeet allotment was completed, the reservation had been making progress toward McFatridge's dreams of a grand cattle yard—a vision increasingly shared by his loose coalition of Blackfeet cattlemen and by off-reservation lessees. The range was already stocked with over twenty thousand head, "but with the excellent and abundant supply of water and grass the country is unsurpassed for the grazing of stock, and with the numerous excellent wild meadows to provide hay for winter feeding, there is no good reason why there should not be at least 100,000 head of cattle on this reservation." Oscar Lipps, the OIA's livestock supervisor, declared: "The country needs the beef, the Indians would like to produce more of it." Lipps suggested a postallotment plan that McFatridge and his coalition embraced: the expansion of a common tribal herd from which cattle could be issued to individual Blackfeet who demonstrated that they could take proper care of the animals. This plan would work in a similar way to the Dawes Allotment Act's system of land patents and competency commissions, with one major difference. Although the Blackfeet's cattle would belong to them in trust, in order to actually sell or otherwise dispose of their animals on the market, individual Blackfeet would need to demonstrate their "competency" as stockmen. Demonstrating competency, in this case, would require that the Blackfeet sell their cattle on the market rather than kill their issued cattle for food.[31]

Lipps also recommended a controversial solution to the problem that the reservation's allotment posed to the agency's cattle-leasing program. On the reservation's unallotted lands, which the OIA planned to put up for sale under the terms of the Allotment Act, Lipps suggested the retention of at least "a few townships of land" that would be set aside for open-range grazing leases. Despite its allotment, McFatridge and Lipps set about maintaining the Blackfeet Reservation as an open range, a clear expression of the OIA's continued intention of incorporating Blackfeet land into the North American beef industry.[32] McFatridge was backed by Blackfeet cattlemen who viewed the sale of the surplus unallotted lands as a means of purchasing additional cattle and as a solution to the problems of off-reservation grazing lessees.

Leading the opposition to this cattle-focused vision of reservation land use and to the sale of surplus unallotted lands was Robert Hamilton, a young Blackfeet whose skills as an organizer eventually seated him at the head of the reorganized Blackfeet Tribal Business Council (BTBC). From this position Hamilton developed access to influential members of the U.S. Congress who went to battle against the interests of the reservation's cattlemen and the OIA. Hamilton was the adopted son of Alfred Hamilton, cofounder of the Fort Whoop-Up whiskey trading post in 1869. Al Hamilton had long proved a headache for the OIA, and during the 1880s he fled to Canada to avoid murder charges. As a descendant of this former whiskey trader, Robert Hamilton was himself a marked man by the Blackfeet agency administration.

The foundation of Hamilton's politics was his desire to achieve Blackfeet economic independence by managing the Blackfeet's land base for the direct benefit of the entire tribe rather than for the OIA and its wealthy stock-owning allies in Browning. The OIA first recognized Hamilton as an obstacle to its control over Blackfeet labor in 1906, when he organized a strike of Blackfeet workers on the Two Medicine irrigation project. The workers struck in October, a critical moment to complete excavations before the ground froze. Within two days and without any violent incidents, Hamilton and his supporters succeeded in negotiating a raise that brought their wages to parity with white workers hired for the project. An infuriated agent Churchill arrested Hamilton and threatened to remove him from the reservation but was denied this authority by his superiors. Explaining the incident to the commissioner of Indian Affairs, Churchill wrote: "The Indians have worked well and have done good work. They have been told so, and a few of them were led to believe that their labor and that of their teams, was equal to that of the white men with larger horses."[33] In fact, the workers struck not because of the agent's praise for their contribution to the Two Medicine project—which would establish an irrigated valley that the OIA would later attempt to sell as unallotted land—but because they were being paid below subsistence-level wages for work that brought them away from their homes and families. The strike's success brought Hamilton to the center of popular politics on the reservation.

The focus of Hamilton's career as a tribal organizer for the next decade would be his opposition to the Allotment Act's proposed sale of the Blackfeet Reservation's unallotted lands. By act of Congress on March 1, 1907, the

Blackfeet Reservation was ordered to be allotted, each enrolled member of the tribe receiving 320 acres. This plan left over 800,000 unallotted acres on the reservation, roughly half of its total area, most of which would be returned to federal domain and sold to settlers under terms proscribed in the original Allotment Act of 1887. Not surprisingly, the reservation's allotment was popular throughout the rest of northern Montana, as the region's cattlemen eyed the possibility of purchasing—possibly for pennies on the dollar—its unallotted grasslands.

Hamilton and his supporters primarily sought to stop the surplus land sale, the most egregious component of an Allotment Act that threatened to dispose of over half the Blackfeet's remaining land base. They eventually achieved this goal not by resisting allotment in its entirety but by organizing the tribe against the act's land sale provisions. At one council meeting in 1909, Hamilton collected the statements of dozens of tribal leaders who objected not to the allotting process but to the sale of unallotted land. Mountain Chief remarked that "in regard to the allotting, when the surplus land is left, I don't want to dispose of this land but I want to hold it for the little children who are coming. This is the strongest idea I have in my head."[34] After twelve years of fighting that stalled any general land sales, Hamilton and his supporters accomplished this goal in 1919, when Congress passed a bill that exempted the Blackfeet from selling any surplus lands and that allotted each enrolled member an additional eighty acres.

Hamilton and his supporters developed relationships with legislators and bureaucrats in Washington DC, who were willing to hear and act on his stories of agency corruption and the shortcomings of the OIA's cattle strategy. Hamilton's first trip to Washington was reluctantly sponsored by the OIA and resulted in an OIA-administered inspection of the agency that, not surprisingly, concluded that the agency was well run and that its administration worked toward the benefit of all Blackfeet. But although his visit was held under the tutelage of the commissioner of Indian Affairs, on his trip Hamilton also met with U.S. senator Harry Lane of Oregon, whom he convinced to lead a separate inspection of the reservation and its management, an inspection that drew very different conclusions—that the majority of Blackfeet lived in a state of destitution, ruled over by an agent and a clique of stock growers more interested in the private acquisition of government-issued cattle than in improving the reservation's broader economic conditions.[35]

McFatridge sought to discredit Hamilton and his supporters by characterizing them as lazy Indians with little personal initiative who survived off agency largesse. McFatridge's particularly vicious enforcement of the OIA's "work or starve" mentality led him to trim the rations rolls that winter, resulting in more cases of starvation. The day after McFatridge explained to the commissioner of Indian Affairs why he ended the distribution of these goods—guaranteed by treaty—to the "able-bodied," he requested authority to spend $1,250 from the same Blackfeet annuity fund to buy himself a new Buick. Upon Hamilton's return to the reservation, McFatridge fumed that "[he] had not worked one single day." Despite the damning evidence from the Lane inspection, McFatridge continued to rule with an iron fist. He answered reports of destitution with a familiar refrain, that Indians, like Hamilton, simply would not work. "There are a number of Indians on the reservation," he explained, "that are able to earn a living who have refused to perform labor for rations for themselves and their families, and I have refused to issue rations to such people. Those who do not have enough to live on," he continued, "can blame themselves." McFatridge sought to establish the value of his own work as agent by representing Hamilton's political efforts as an illegitimate form of labor.[36]

The next year, Hamilton tried to organize a second trip to Washington, which brought him into a series of direct clashes with McFatridge. Over the winter, Hamilton had organized a BTBC meeting without notifying McFatridge. The council voted to send Hamilton back to Washington along with two tribal elders, Wolf Plume and Young Man Chief, who would help finance the trip. After discovering the plan, McFatridge was enraged, describing to his superiors that Hamilton was "simply an agitator and has caused more trouble in this tribe of Indians than any other person." McFatridge ordered the tribal police after the trio and had them arrested at the Browning railroad station. After their release, they tried a second time, riding off the reservation to pick up the Great Northern Railroad at Cut Bank. McFatridge gave chase and once again arrested them. But the

OIA could no longer overlook the agent's constant overreaches of authority, and it fired McFatridge later that month. He fled to Alberta with $1,200 of tribal funds, cash raised from the annual grazing tax.³⁷

McFatridge's sudden departure left the agency administration in disarray. From 1916 until 1920 seven different OIA employees managed the agency. During these years, the center of power on the reservation shifted from the OIA to Hamilton and the BTBC, his governing body, which stalled the sale of unallotted lands and pressured Congress to overturn the provision altogether. Hamilton and his supporters sought to use allotment to their advantage to secure a victory against the reservation's stock-growing elite.

Key to their success was their relationships with emerging radical centers of power in both Montana and federal politics, including the Nonpartisan League. In 1918 the tribe selected its former allotment agent, Louis S. Irvin, to serve as its tribal attorney.³⁸ Irvin was himself married to a Blackfeet woman. He quickly sided with Hamilton's faction against the OIA's flawed plans to sell oil leases on the reservation. In 1920 Irvin served as Burton K. Wheeler's running mate in Montana's gubernatorial election.³⁹ Although Irvin and "Bolshevik Burt's" run on the Nonpartisan League ticket was an electoral disaster, the campaign catapulted Wheeler into the U.S. Senate as a Montana Democrat two years later, during the height of the northern plains' regional economic depression. By the 1930s Wheeler was chair of the Senate Committee on Indian Affairs and the primary sponsor of the Indian Reorganization Act, which the Blackfeet were one of the first tribes to sign.

Although unsuccessful in reversing OIA policies regarding the stock industry, the reservation's anticattle faction took the lead in the Blackfeet's effective resistance to the Allotment Act's provision for surplus land sales. These debates revealed conflicting visions of how the reservation's land and resources would be best utilized, as well as tensions over understandings of labor and selfhood that underlay the OIA's assimilation project and its reception by enrolled Blackfeet.

The Piegan Farming and Livestock Association

Historians have been right to criticize federal allotment policies as a means of dispossessing American Indians of their lands and labors. Tom Wessel has described the allotment of the Blackfeet Reservation in particularly stark terms: "Instead of independent agricultural communities, the government created pockets of rural poverty physically fractionalized and politically factionalized."[40] It is no surprise that Eloise Cobell and other Blackfeet have been national leaders in ongoing attempts to reform the management of Indian trust assets, as the crippling legacies of allotment and fractionation remain evident across the Blackfeet Reservation. However, in the immediate wake of allotment, the Blackfeet did not give in to the colonial prerogatives that would supposedly be accomplished by the privatization of their tribal lands. As Emily Greenwald has argued in her study of allotment policy, many Indians ironically used their allotments to "remake themselves as Indians in the context of a policy that sought to destroy Indianness."[41] In a recent study of Creek history, David Chang has remarkably shown how allotment policy led not to assimilation but to a complex formulation of Creek racial nationalisms during the late nineteenth and early twentieth centuries.[42] Far from signaling the disappearance of Blackfeet understandings of land and livelihood, allotment opened the door for an interwar renaissance of Blackfeet environmental knowledge. Through the development of the Piegan Farming and Livestock Association (PFLA), the Blackfeet accommodated allotment to Blackfeet notions of áuasini.

Even a colonial institution like allotment could be reworked by the Blackfeet to fit within their broader communal kinship traditions. As the radical positions of Robert Hamilton and his supporters indicated, many Blackfeet saw in their individual allotments opportunities to reconnect their individual labor back to a broader communal tradition of value creation, one that subverted modern capitalist understandings of discrete individual selfhood. Through the PFLA, Blackfeet sought to use their individual allotments to develop a communal subsistence economy on the reservation. While it is too much to suggest that allotment offered a powerful mode of resistance to colonization—after all, sales of patented allotments and fractionation have subsequently devastated the Blackfeet's land base—by the mid-1920s many Blackfeet had nevertheless utilized their allotments to develop effective cooperative organizations that blended subsistence and market agriculture in a way that resonated with Blackfeet understandings of place, sustenance, and selfhood.

The PFLA emerged during the fall of 1921 as a reservation-wide Blackfeet organization committed to reaching the subsistence goals of Campbell's Five-Year Industrial Plan (FYIP). Adopted in March 1921, the FYIP aggressively focused agency energies around the task of expanding subsistence agriculture on the reservation, with the main objective of achieving Blackfeet agricultural self-sufficiency by 1926. Although the plan did place its faith in a kind of mixed agriculture sure to conjure images of white yeoman farmers, the program's emphasis on communal self-sufficiency nevertheless signaled a radical departure from the OIA's traditional emphasis on assimilation. Paired with the Blackfeet-organized PFLA, the FYIP-PFLA achieved widespread popularity with the Blackfeet. For his part, Campbell became the OIA's brightest star; by 1923 he spent most of his time away from Browning, touring other reservations throughout the United States and Canada, giving lectures, and trying to convince other agency administrators to adopt his program, a task in which he was mostly unsuccessful. Hamilton's Blackfeet Tribal Business Council fired Campbell in 1928, claiming that his administration was no longer needed. Reinstated by the commissioner of Indian Affairs—the BTBC's legal authority to fire agency personnel being questionable—Campbell's tenure boiled into a controversy over Blackfeet sovereignty and OIA administration that led directly to the desks of Senator Wheeler and, eventually, John Collier, authors of the Indian Reorganization Act six years later.[43]

Year 1 of the FYIP set fairly modest goals. Sending out a wide call for participation, Campbell offered all enrolled Blackfeet the necessary fencing materials to enclose a forty-acre field on their allotment. Each family that built fences also received one milk cow, twenty sheep, and twelve chickens. Using agency funds, Campbell also built a flour mill in the Heart Butte district, optimistically awaiting the reservation's first major harvests of grain. More significantly, Campbell visited every household on the reservation that summer, along with an interpreter, the agency doctor, its farmer, and its livestock supervisor. He explained the program personally, encouraged Blackfeet participation, wrote reports, and photographed the reservation's population outside of Browning, where previous agents had seldom ventured.

Because of the way it engaged Blackfeet understandings of relational selfhood, this personal attention during the summer of 1921 may have been one of the keys to the FYIP-PFLA's success. The Blackfeet concept of "medicine power" had been invoked earlier by OIA assimilationists to explain Blackfeet reluctance to pursue agriculture. According to this understanding of power, individuals needed to borrow or beg for abilities with which they were not originally endowed. Rather than personal initiative amounting to the basis for individual success—one of the base truisms of Anglo-American autonomous selfhood—Blackfeet selfhood, *mokaksin*, constituted an assemblage of personal capabilities gathered from other human and nonhuman persons. In the prereservation era, Blackfeet commonly forged these relationships through dreams, ritual fasts, sweats, and self-torture. Sometimes the powers derived could be embodied within collections of objects—medicine bundles—and then exchanged with others. The animal world was the original source of ability and knowledge. Animals held powers in their pure forms; humans drew on a panoply of watered-down powers granted them by their nonhuman patrons.[44]

Although the ritual practices associated with medicine power, particularly self-torture, had been severely curtailed by OIA administrators from the 1880s onward, reservation Blackfeet continued to organize their senses of selfhood around the concept. In 1923, for instance, the Blackfeet adopted Tom Gibbons, Jack Dempsey's challenger in the title fight for heavyweight boxing held that summer in Shelby, Montana, just east of Browning on the Great Northern Railroad. They granted Gibbons the name Pony Kick in Hands to describe his unique boxing abilities, surely derived from his kinship with horses. Despite losing to Dempsey in fifteen rounds, Ponykick-in-hands remained a folk hero on the reservation and a testament to the modernization of medicine power in Blackfeet popular thought by the 1920s. Campbell's house-by-house canvass of the reservation thus fell within an established Blackfeet framework of medicine power, ritual, and relational selfhood that undoubtedly encouraged the success of the FYIP-PFLA. Through these personal interactions, Campbell transferred the agricultural abilities that his medicine powers represented, increasing the share of relational selfhood accessible by PFLA members.

According to Campbell's original plan, the main goal of the FYIP's second year would be to "discourage the nomadic habits and inclinations and to create for the [participants] an incentive for staying at home and strictly attending to their industrial activities." "Nomadic habits" was a euphemism for the Sun Dance, the annual Blackfeet gathering that agency authorities had tried to suppress for over thirty years. Campbell also opposed the Sun Dance and other large tribal gatherings, which were bastions of traditionalism, as well as the main setting for ritual self-torture, because they took the Blackfeet away from their farms and ranches, sometimes for a month or longer. Other agents had tried to outlaw the Sun Dance without success. About ten years earlier, McFatridge tried to merge it, without success, with an annual Fourth of July celebration in conjunction with his assimilation campaign. The establishment of the PFLA provided a workable solution to the Sun Dance dilemma for Campbell.

The third year rolled out the FYIP's most revolutionary goal: "elimination of the necessity for the Indian to leave his home in search of employment and to help him work out a plan whereby he may spend a three hundred sixty-five day working year on his own allotment." Unlike three decades of OIA programs that sought to subordinate and incorporate Blackfeet labor into Montana's livestock industry, the FYIP emphasized the goal of eschewing Blackfeet wage labor altogether. More than anything else, this goal of the FYIP provided a complementary element to Blackfeet understandings of food and place. The land and its interrelationships could provide a livelihood for those who dwelt upon it.

But it would be a mistake to characterize the FYIP and the PFLA as being entirely subsistence based. An auxiliary goal of the organization was to "work out a system of marketing whereby the Indian can place upon the open market any surplus farm or live stock products." The PFLA was not a retreat from the market economy but a program to establish Blackfeet means of production that could accommodate the reservation's subsistence requirements and also accumulate capital. Accordingly, the program included the goal of 75 percent of allotments producing at least ten acres of wheat by year 4, enough to feed the reservation, but also by year 5 the goal of 70 percent of the reservation's allotments producing an agricultural surplus for sale, either grain, vegetable crops, livestock, or hay, which was probably the most common surplus commodity.[45]

4. White Grass Chapter PFLA, ca. 1924. Courtesy of William Farr.

During the first summer, the Blackfeet developed twenty-nine PFLA chapters to administer the project, each chapter loosely organized around existing kinship relationships. About fifteen different families, most of which were interrelated through marriage, comprised each chapter. This familial geography was the legacy of a previous relocation effort. In 1902 agent James Monteath relocated a substantial number of full-blood families to Heart Butte and other southern districts in anticipation of irrigation projects.[46] When the reservation was allotted between 1907 and 1912, most of these neighbors were allotted together, "so that a chapter membership [of the PFLA]," wrote a reporter for the *Indian Leader* in 1923, "in many instances comprises a large family group."[47] Organized around existing family units, PFLA chapters existed similar to band units, drawing on prereservation Blackfeet social structure (see figure 4).

One major way that PFLA chapters formed a cooperative organization was through group purchases and sharing equipment. All twenty-nine chapters shared the same flour mill, constructed in 1921. They also shared two grain threshers that traveled across the reservation during harvest season. But most chapters bought other machinery for use by their own chapter members, including hay binders and rakes and other machines. The *Indian Leader* also reported on this phenomenon: "This farming machinery is turned over to the president of the chapter and he is held responsible for its housing and safe-keeping and the various chapters all over the reservation are building community machinery sheds for this purpose." The cooperative use of machinery helped PFLA members avoid the large individual debts faced by regional farmers forced to purchase equipment on their own.[48]

The PFLA's cooperative mission extended beyond the bounds of financing machinery purchases, however. One of the PFLA's strongest admirers, General H. L. Scott, who inspected the reservation for the Indian Office in 1925, lauded the program through the story of George Wren, an older chapter member who suffered from blindness and was unable to plant in 1924. "The other members of his chapter, noting his condition," wrote Scott, "got together and put in his crop in addition to their own, a circumstance which would not have happened had there not been a chapter organized."[49] Belonging to the PFLA chapter expressed a series of kinship obligations that extended to providing labor beyond the boundaries of one's individual allotment.

Even after the first year, the PFLA's results impressed commentators. The old-time Indian office inspector Oscar Lipps visited the reservation in the fall of 1921. He reported on one Blackfeet named, ironically, After Buffalo. "The old man and wife and two grandsons live together," he remarked. "They raised nothing last year except a few potatoes. This year they raised 40 bushels of wheat, 30 bushels of oats, 50 sacks of potatoes, some rutabagas, beets, carrots, peas, corn, etc. They put up hay, sold 20 tons and have one stack left."[50] With hay selling for around $12 a ton, After Buffalo and his family not only grew enough food to sustain themselves for an entire year but also generated a cash income of over $200, a fortune by early twentieth-century reservation standards. Satisfied with the program's results, Lipps ended his report: "Thus stands the record of the efforts of these old buffalo-eaters and hunter-sportsmen in their first attempt at subsistence farming." Even a veteran representative of the OIA grudgingly admitted the PFLA's success. In the fall of 1921 the Blackfeet harvested and milled eleven hundred bushels of grain. The following year they milled fifteen thousand bushels, enough to feed the reservation.[51]

Nevertheless, the PFLA was not without controversy. An older generation of assimilationists denounced the program from the start as an unrealistic experiment in agriculture that was "doomed to failure," as James Willard Schultz put it. Schultz, a white settler on the reservation in the late 1800s and the popular author of *My Life as an Indian*, a book that described his marriage to a Blackfeet woman and their life in Browning, published a scathing editorial on the PFLA from his retirement home in southern California in 1921. On the reservation, Schultz was an advocate of the cattle regime that had bankrupted the Blackfeet during the 1910s. He published the editorial under the title "The Starving Blackfeet," drawing upon a long-standing tradition of similar charges against the OIA that reflected a Progressive Era white philanthropic mission to lessen the suffering of American Indians. Except this time, the charges of starving Blackfeet were unfounded. Schultz represented the Blackfeet as "starving" under the PFLA to help justify a return to previous administration tactics that focused on procuring livestock for the reservation's previously well-off stock growers. Schultz did not advocate the growth of Blackfeet communal self-sufficiency but instead worked as an apologist for the failed policies of individuation and assimilation that the PFLA overturned. He ended his editorial with one last rebuke for the project: "There remains but one hope for the Montana Blackfeet, and that is, to obtain from the government a portion of the value of the vast territory arbitrarily taken from them by presidential executive orders."[52] Although Schultz was right to request compensation for lands stolen from the Blackfeet during the nineteenth century, his opposition to the PFLA revealed his shortsightedness in retreating toward the prospect of "trading land for food," the old tactic that he believed would spare the Blackfeet from colonial annihilation but that could not ensure their long-term prosperity.

Other critics decried the project as Bolshevist and communistic. These charges mostly came from men like Schultz who continued to cling to the promise of an individuated cattle economy. Supporters of plans to sell oil and mining leases on the reservation also resorted to harnessing the Red Scare to support their political mission. In a report written in 1924, General Scott responded to these criticisms, retorting that "the program is in no way communistic, it works for individual effort in a cooperative way." He further commented that the PFLA was on the leading edge of other cooperative associations forming throughout the United States and Canada in response to the challenges of industrial agriculture: "The white farmers are forming cooperative societies all over the country for mutual support and benefit as many other industries are organized."[53] Scott realized that the PFLA was situated within a broader Progressive movement not toward a radical communism but toward a cooperative social democracy based on understandings of social selfhood rather than pioneer individualism that perhaps reached full expression during the 1930s.

In this sense, the PFLA drew on Blackfeet traditions of relational selfhood to inform its cooperative endeavors, but it would be a mistake to characterize the program as traditional. The PFLA was significant because it provided a means to blend Blackfeet knowledge of the world with nontraditional forms of food and market production. Unlike previous OIA programs, its goal was not the assimilation and incorporation of the Blackfeet into the American body politic as colonial subjects but the establishment of Blackfeet communal self-sufficiency by partnering modern agricultural procedures with an organization that drew on Blackfeet understandings of selfhood, labor, and the origins of value.

Conclusion

While the allotment of the Blackfeet Reservation stood to further incorporate the Blackfeet as unequal members of a colonial society, Blackfeet efforts to integrate conceptions of áuasini into the privatization of the reservation landscape helped to establish a cooperative mode of subsistence with the development of the PFLA. Throughout the allotment process, the Blackfeet battled among themselves and the OIA largely along political lines drawn by the cattle industry. By utilizing the discourse of predation and production as a language to track the relative progress of Blackfeet assimilation, the OIA dissociated the nonhuman labors necessary to livelihood from the private labors of individual Blackfeet. The reservation stock industry provided a suitable political ecology for making these social distinctions. Rather than conceptualizing cattle issued to the tribe as public elements of the reservation's broader environment, the OIA and its Blackfeet allies represented cattle as forms of animal capital whose value accrued to the work of their owners.

By privatizing tribal lands, the allotment of the Blackfeet Reservation also served to individualize Blackfeet relationships to the nonhuman environment. But through the cooperative subsistence agriculture practiced by the PFLA, the Blackfeet developed a new mode of livelihood that constructed a series of broader environmental and social relationships atop the individuated grid of private property that allotment had established. By reintegrating indigenous concepts of human-nonhuman kinship into reservation agriculture, the Blackfeet established an alternative mode of livelihood that expanded their economic options as a colonized society.

By the early twentieth century, Anglo-American understandings of predation and production had come to hinge on concepts of individual and social selfhood. Representing oneself as a producer required a disavowal of the nonhuman labor necessary for the creation and distribution of value and capital. Subsuming the animal labors of cattle under the individual self of a stock grower was a major process in which this representation of capital worked out on the ground. Representing non-stock-related forms of work as nonlabor or even as predation, the OIA and its stock-growing allies on the reservation sought to reorient Blackfeet labor to serve the broader goals of the stock industry's capital accumulation. As the next chapter demonstrates, the logic of predation and production, and its relationship to Anglo-American concepts of selfhood, helped structure historical narratives of the frontcountry that made it a landscape capable of both conservation and exploitation. Colonialism's disavowal of its own exploitative labors relied on the ongoing fabrication of predatory persons, both human and nonhuman. Under the cultural and ecological regimes of the capitalist livestock industry, it remained necessary to designate predators in order to establish oneself as a producer.

5

Unnatural Hunger

On New Year's Eve of 1900, diners across the United States ate twelve bison from northern Montana. Raised on the Flathead Reservation by Charles Allard and Michel Pablo, the animals were slaughtered before fifteen hundred spectators at the Montana State Fairgrounds in Helena, then shipped by rail to restaurants in St. Louis, New Orleans, San Francisco, and New York. The meat was "coarse and dry" compared to the lamb chops and rib roasts preferred by American epicures at the dawn of the century, but the bison whetted romantic appetites for the nation's frontier past. Hunger for bison, remarked the *Minneapolis Tribune*, "will never die as long as the memory of man runs to the era of the conquering of the west."[1] In tendering communion with a century of conquest, the bison's flesh connected a nation's elite to the landscapes and labors of its colonizing forebears.

Never fully vanquished from the North American plains, bison began to feed a broader set of eaters during the Progressive Era. From affluent gourmands to welfare recipients, bison meat ended up in public relief houses, at Indian agencies, and in the dining cars of railroad tourists bound for Glacier and Banff. At first glance, this taste for bison seems somewhat unexpected given the animal's status as a major object of wildlife conservation during the early twentieth century. But by the 1910s two of the era's flagship bison conservation projects, the Montana National Bison Range and the Wainwright Buffalo National Park in Alberta, quickly became breeding grounds of bison overpopulation, and the surplus animals increasingly wound up in American and Canadian kitchens.

Established by both the U.S. and Dominion governments in 1908, these two refuges functioned primarily as ranches for bison and, at times, other ungulates. Conservationists and federal authorities envisioned and managed the parks as closed settings free from the effects of human and nonhuman predators. Bison herds at both locations quickly reached unsustainable sizes, and refuge administrators began slaughtering the surplus animals. Planned and administered as ranches for supposedly wild ungulates, these national bison refuges represented the saturation of colonial idioms of predation and production within the broader goals of Anglo-American wildlife conservation. Their history reveals how local understandings of human and nonhuman labor shaped national frameworks of colonialism and human-animal relationships during the early twentieth century.

The dissociation of predation from production structured efforts to preserve wildlife in Montana and Alberta. In the first decades of the twentieth century, conservationists bridged the political and social gaps between eastern animal preservationists and the interests of western livestock growers by ascribing the near extermination of bison and antelope to overhunting rather than habitat loss, casting blame on the so-called predators, such as wolves and hunters, rather than the so-called producers, the stock growers who had transformed most existing bison and antelope habitat into rangeland for cattle and sheep by the end of the nineteenth century. Western cattlemen and other powerful stakeholders forced eastern philanthropists to frame their conservation policies within a colonial narrative of predation that has broadly informed regional wildlife controversies ever since.[2] In the case of bison conservation, William Hornaday, the founder of the American Bison Society (ABS), was himself largely responsible for crafting and articulating this narrative. Over the course of a fifty-year span, Hornaday published various histories of both bison extermination and bison recovery that denigrated the predators and celebrated the producers. More broadly, Hornaday and other eastern conservationists drew upon local historical narratives that characterized wolves, Indians, and Basque sheepherders as "predators" who preyed upon the West's productive colonial institutions, primarily the livestock industry, but also the region's game refuges. These narratives fell upon a receptive western public eager to displace the environmental effects of colonization and industrial agriculture onto the presupposed actions of bloodthirsty hunters—both human and nonhuman. The same narrative of predation was equally attractive for wider national publics that began celebrating their western heritages during the early twentieth century, the easterners embarking on tours of Glacier, Banff, and other sublime landscapes of the Northern Rockies. Conservationists would gain more local cooperation if they could draw on these narratives and assign blame for conservation problems to actors already locally disenfranchised as predatory while casting powerful white cattlemen, settlers, and other stakeholders as productive.

In these ways, dissociating predators from producers provided conservationists with a necessary cultural procedure for bridging the divide separating western and eastern conservation politics during the Progressive Era.[3] Bison conservation and repatriation in Montana was a central focus for the Progressive Era conservationists and a pet project of Theodore Roosevelt himself, who was both a friend of Hornaday and a member of the American Bison Society. From the beginning of the ABS's fund-raising efforts, Hornaday and his East Coast lieutenants worked to establish themselves as producers, seeking to overcome western suspicions of their sentimental, preservationist impulses. Through strident predator management policies and rhetorical thunder, the ABS leveraged its existing regional alliances and forged new relationships with centers of western power, the livestock industry in particular. By framing themselves as producers of wildlife, members of the ABS sought to link their eastern politics of preservation with a more local and western politics of economic development. Prominent conservationists and members of the ABS, such as Madison Grant, William Hornaday, and Edmund Seymour, all based in New York, adapted their conservation prerogatives to understandings of predation and production that had emerged in the Northern Rockies frontcountry over the previous four decades.

Adapting to western land-use practices by killing predators was a centerpiece of this conservation paradigm. Maligning predators offered an expedient route toward identifying shared concepts of labor, value, and profit that rested at the heart of both eastern and western political outlooks. Identifying the predatory habits of wolves, Indians, and Basques as the primary threats facing bison and antelope populations, the ABS and regional stock growers overlooked the larger environmental effects of the borderland's forty-year transition into a colonial landscape of homesteads and cattle ranches. Although swollen herd sizes and inadequate grassland habitat posed the greatest challenge to the management of Wainwright Buffalo Park and the National Bison Range, their administrators maintained a focus on eliminating predators through the 1920s and 1930s. These efforts brought conservationists further into alliance with the region's livestock industry. By utilizing a logic of predation that supported capitalist stock raising as natural and productive, organizations like the ABS succeeded in mobilizing regional support for wildlife refuges, as well as popularizing the by-now-common truism that predation was the main impediment to successful wildlife conservation.

By 1931 the U.S. Animal Damage Control Act enacted these conceptions into federal law. Department of Agriculture hunters had effectively exterminated wolves from Montana five years earlier. Just as the bounty bills discussed in chapter 2 had proved more effective for hardening the logic of predation into official colonial discourse than as tools for managing wolf behaviors, the Animal Damage Control Act reflected more the institutionalization of the logic of predation into wildlife conservation than either the political or ecological realities of animal preservation. By cultivating a common ground that represented their own labors as productive, both western ranchers and eastern wildlife conservationists found something useful in the borderland's colonial narratives of predation. Moreover, these local constructions of predation and production migrated eastward during the 1920s to influence conservation policies on a federal level, providing a natural-historical narrative of immutable predator-prey relationships that shaped legislative debates over the Animal Damage Control Act.

Positioned at the heart of controversies over this bill were local western understandings of what constituted categories of legitimate labor and "natural hunger," notions based on a colonial history of dissociating predation from production. Even those contesting the 1931 Animal Damage Control Act, like the American Society of Mammalogists (ASM), took these understandings for granted. In private correspondence to Paul Redington, chief of the Bureau of Biological Survey, the Johns Hopkins mammalogist A. Brazier Howell conceded that the ASM's admonitions against predator control did not extend to wolves, "who were truly killers."[4] Stanley Young, chief of the U.S. Bureau of Biological Survey's Division of Predatory Animal and Rodent Control, was happy to hear this universal prejudice echo from the halls of science. "The animal is one hundred percent criminal," he retorted in a letter to his friend, the wilderness advocate Arthur Carhart, "more often killing to satisfy his lust than to satisfy a natural and reasonable hunger."[5] Narratives of predation demarcated the boundaries between acceptable and unacceptable modes of exploiting human and animal flesh. Only by integrating these local colonial narratives into the core tenets of Anglo-American wildlife politics could federal conservation projects succeed on western terrain. Progressive Era conservation mattered because it established a legal validation backed by scientific authority of the disenfranchisement of those human and animal livelihoods deemed unproductive under colonialism and capitalism. Dissociating predation from production was necessary to the conservationist's work of repatriating bison and eradicating wolves.

Narratives of Predation and the Origins of Bison Conservation

In 1909 William Hornaday celebrated the founding of the Montana National Bison Range by serving as the refuge's first chronicler. Eager to commemorate his own contributions to the bison's restoration, he exaggerated the project's difficulties, beginning his account by justifying "the privilege of writing [the park's] history" as "compensation for the labor performed" by himself and other prominent members of the ABS.[6] From the very moment of the park's founding, a particular logic of labor as production rather than predation framed narratives of the park's inception. This understanding situated the work of bison conservation as a productive effort that reversed the historical forces of predatory overhunting. Hornaday and the ABS utilized this logic to cast their own labor as conservationists as a productive force within the broader setting of western economic growth.

In his official recollection of the range's history, Hornaday paid little attention to the significant regional and national opposition that faced the ABS's Montana plan. Opponents of the National Bison Range had criticized the refuge as a regression of the American West's modern history. The *Washington Post* explained its opinion of the bison project in evolutionary terms that stressed the natural succession of predation with production: "The wolf and the coyote must give place to the collie, the pointer, and the setter, just as the Indian receded before the white man." The bison "had his day," continued the editorial. "In our civilization there is no place for him. We demand the milk, butter, and cheese cow and the beef bullock."[7] A special report on the bison range from the *Minneapolis Morning Tribune* opined that the preserve seemed a "queer turning back of history to the old settlers of the northern plains."[8] Linking the slaughter of bison as a stage in the process of civilizing the Montana plains, opponents of the project likened bison conservation as a step backward within the state's progressive historical narrative.

Finding a place for bison conservation within this matrix of progress and production was a major public relations challenge for Hornaday and members of the ABS. But Hornaday emphasized other challenges. For one, there was nothing novel about the National Bison Range. Small herds of bison already existed on public lands, such as the herds in Yellowstone National Park and the Wichita Mountains in southwestern Oklahoma. The ABS itself had actually stocked the Oklahoma range with fifteen bison from the Bronx Zoo in 1907. And above the forty-ninth parallel, Canada had already beaten the ABS in establishing a bison herd on the northern plains. In 1908 the Canadian government established Buffalo National Park in Wainwright, Alberta, a bison conservation project with over four hundred animals, dwarfing the Bison Society's proposed herd of fewer than one hundred. Stock growers were understandably vexed by Hornaday's call to establish another bison refuge, since it seemed that other public initiatives had already succeeded in protecting the animals. Creating more refuges would merely mean more public land set aside for bison at the expense of private grazing leases. "Shed no tears over the bison," claimed one editorial. "Keep him as a curiosity. That is all he is good for."⁹ From the standpoint of existing conservation projects and regional politics, the National Bison Range was ill conceived and poorly timed, and under the current circumstances, finding land for the project would not have been easy.

The ABS, however, found a windfall in 1908, following the allotment of the Flathead Reservation. An act of Congress, introduced by Senator Joseph Dixon of Missoula, had allotted the reservation in 1904, splitting 228,434 acres among 2,390 original allotments registered to individual Salish and Kootenai tribal members. Four years later, nearly one million acres of reservation land remained unassigned and were classified as surplus, stripped from tribal ownership, and opened for public sale.¹⁰ The ABS was one of the buyers. Established later that spring, the National Bison Range encompassed eighteen thousand acres of these former tribal lands.

The privatization of the Flathead Reservation had provided a key moment of opportunity for Hornaday and the ABS. One year earlier, in 1907, Hornaday had enlisted Professor Morton J. Elrod, chair of the University of Montana's Biology Department, to conduct a survey of the reservation's surplus lands and locate range for the proposed refuge. Elrod's job was made relatively easy by the fact that, over the previous twenty years, two Native men—Michel Pablo and Charles Allard—had grazed a herd of bison on the southern third of the reservation, near the foothills of the Mission Range. The Pablo herd had descended from a handful of bison driven from across the mountains in 1878 by Samuel Walking Coyote, a Blackfeet man who had married a Salish-Kootenai woman.¹¹ Less than fifty miles from Missoula, the land the bison grazed was located near the town of Dixon, recently named for the reservation-opening senator by his grateful business associates. Not only was this good year-round pasture, but over the past twenty years it had proven an ability to support herds of over five hundred bison. Elrod submitted his report to the ABS, which referred it to the Senate Indian Affairs Committee early in the spring of 1908. As chairman of the Senate Indian Affairs Committee, Dixon sponsored legislation that passed the Senate just in time for incorporation into the House's Agricultural Appropriation bill, which President Roosevelt signed into law on May 23, 1908. The bill authorized a budget of $30,000 to purchase Flathead Reservation land from the Office of Indian Affairs for the National Bison Range. Without Indian allotment, the National Bison Range would not exist.

Hornaday credited his own business savvy for the range's establishment. He acknowledged the challenge of finding land for the refuge without "interfering with the settlement of the country."[12] But rather than stating how the project's success relied on Congress's dispossession of Native land, Hornaday instead emphasized the Bison Society's skill in political lobbying. Finding land for the range, remarked Hornaday, "was done with the same briskness and precision with which the best-managed business corporation takes up and acts upon an important matter when the urgency for action is very great."[13] Using this commercial language, Hornaday represented the society's actions as productive in the western colonial sense, eliding the expropriation of Native land and excluding the Salish-Kootenai contribution to bison conservation.

After locating land for the National Bison Range, finding bison to stock the refuge became Hornaday's second major task. Like the rangeland itself, the project's bison had Indian origins. With allotment imminent, Michel Pablo, the surviving owner of the Flathead bison herd, had put his animals up for sale. The Canadian government, which had created a similar bison preserve near Wainwright, Alberta, was also in the market for bison. Like the National Bison Range, the mandate of Canada's Buffalo National Park required that it be stocked with pure-bred bison, not with cattalo, beefalo, or other hybrid species developed during the era by regional entrepreneurs.[14] Numbering over five hundred head, Pablo's herd was the largest remnant of plains bison in the world. The Dominion government and the ABS developed an intense rivalry as they competed to acquire the herd.

Sensing trouble with Hornaday, Pablo sold the herd to the Canadians. Like many others on the Flathead Reservation, Pablo had faced an antagonistic relationship with the U.S. government during the allotment process. Although he stood to potentially profit from the sale of the bison herd, he was ultimately forced to sell the animals because of the allotment act. With privatization and the disappearance of tribally owned land, Pablo was left with an eighty-acre parcel on which he could no longer graze the herd without cobbling together extensive grazing leases either from the Office of Indian Affairs or from other allottees at great expense. Moreover, the transfer of the bison range from tribal authority to the federal government seemed to indicate the Americans' intentions of seizing control over the animals if they remained on public land. These circumstances facilitated the Canadians' negotiations with Pablo. Left with few other choices, Pablo sold the herd to the Wainwright Buffalo National Park for $200 a head shortly after Congress authorized the National Bison Range in 1908.

Following the sale of Pablo's herd, an infuriated Hornaday was left to locate other options for stocking the National Bison Range. In personal correspondence to ABS members, he fumed over Pablo as a "half-breed Mexican Flathead" who sold out the nation's stake in bison to the Canadian government.[15] By piecing together a much smaller herd from individual animals scattered across the United States, Hornaday thought he could succeed without Canadian or Flathead assistance. Over the summer of 1908, he arranged to purchase forty bison from Alicia Conrad in Kalispell, widow of Joseph Conrad, a Montana business magnate and former Fort Benton whiskey trader. Conrad had purchased these bison from Charles Allard in the late nineteenth century. Hornaday also contacted James J. Hill in St. Paul, owner of the Great Northern Railroad, who agreed to donate three bison bulls he kept at his country estate in North Oaks, Minnesota—though these bison, in separate incidents, were killed both by his neighbors and by his son before they could be delivered to the Bison Range. Nevertheless, by the end of 1908, Hornaday had made sufficient arrangements to acquire a nucleus herd for the range.[16]

Hornaday represented his search for a bison herd as a paragon of productive labor, a form of value creation forged in opposition to the predation of Indians and others. "In these days of destruction," he recorded, "any man's interest in wild life can be measured by the amount in cash, and hours of labor, that he annually expends in the promotion of measures for wild-life protection."[17] Hornaday situated wildlife conservation as a series of economic transactions premised on growing the stocks of certain wild animals. In essence, the task of a conservationist was similar to that of the cattleman: protecting the herd from predatory agents to secure its increase.

Hornaday supplemented conservation's productive ethos with narratives of predation that represented the decline of wildlife populations as a consequence of overhunting, narratives that grew intertwined with common understandings of American environmental history. The near extermination of bison offered one of the most dramatic examples of this predatory narrative in North America, even the world. Hornaday himself authored one of the first iterations of this narrative. He began his lifelong public association with bison while employed as the Smithsonian Institution's taxidermist during the 1880s. In 1886 Hornaday had convinced the Smithsonian to send him and a field outfit to Montana to collect and taxidermy specimens for the museum. Following his trip to the West, Hornaday wrote the first natural history of American bison, The *Extermination of the American Bison*, in 1887.[18] Hornaday emphasized the complicity of hide hunters and sport hunters in the bison's wanton destruction. Taking a romantic view of American Indians, Hornaday also leveled contempt for the corrupting influences of the market economy on Native hunters. However, throughout the book, Hornaday was careful not to criticize the colonial process itself. Accepting the inevitability of western conquest, Hornaday merely objected to his fellow colonialists' seeming lack of interest in wildlife preservation.

Hornaday detested the practices of other hunters, from sportsmen to subsistence gatherers, but he had a voracious appetite for those same animals himself. Although he developed a reputation as a virulent antihunter, Hornaday was himself responsible for killing some of last free-ranging bison in North America. On his expedition to Montana in 1886, Hornaday shot six bison in the Judith Basin to taxidermy for the Smithsonian.[19] This was typical behavior for Hornaday. As director of the Bronx Zoo, he later contracted the notable wild animal trader Carl Hagenbeck to procure four Indian rhinoceroses in 1902. Hornaday was undeterred when Hagenbeck's team killed forty rhinos in the process of supplying the zoo's order for live animals. Of the three that survived transport to New York, Hornaday wrote that "they will be of more benefit to the world at large than would the forty rhinoceroses running wild in the jungles of Nepal, seen only at rare intervals by a few ignorant natives."[20]

Likewise, during his career at the Bronx Zoo, Hornaday founded the park's National Collection of Heads and Horns, a clear continuation of his earlier interests in the material capital of dead animals. In 1909 he tried to talk Fred Kennard into traveling to Newfoundland on a similar mission to kill and retrieve caribou heads. "I long most ardently," he intimated, for "about four big heads of caribou, such as were the largest to be secured." "Ordinary heads," he added, "will not do."[21] Hornaday similarly sought the heads of mountain goats from Banff, enlisting the help of a local guide, James Simpson, a man who lost his hunting license five years later for poaching.[22]

For Hornaday and other conservationists of his class and temperament, such as Theodore Roosevelt, hunting was an acceptable activity so long as it maintained a dissociation from predation. These men did not kill to eat, as Karl Jacoby and others have demonstrated in their histories of American sport hunters in the Progressive Era, and were generally suspicious of lower-class men who did.[23] In his book *Our Vanishing Wildlife*, Hornaday expressed these sentiments in his descriptions of immigrant "pothunters," claiming that "every foreigner who sails past the statue on Bedloe's Island and lands at our liberty-ridden shore . . . buys a gun and goes out to shoot free game."[24] Moreover, Hornaday's xenophobia cast the stakes of civilization alongside crusades to control predatory beasts, publishing in the annual reports of the Bronx Zoo that "unless man is willing to accept a place in the list of predatory animals which have no other thought than the wolfish instinct to slay every living species save their own, he is bound by the unwritten laws of civilization to protect from annihilation the beasts and birds that still beautify the earth."[25] Using narratives of predation to mark destruction as a characteristic of the uncivilized, Hornaday unconsciously looked to resolve the paradoxes of conservation's own reliance on animal death. Crafting a narrative of predation that cast subsistence and even market hunting as inherently uncivilized, Hornaday sought to justify his own predaceous practices on the grounds of their scientific and philanthropic benefit. Ultimately, this effort hinged on broader social understandings of production, framed in opposition to predation, that had developed in the western borderland.

Crafting the Narrative of Productivity

East Meets West: Enlisting Easterners

Tensions over bison conservation often emerged along regional lines. Eastern conservationists sought to save bison and other wildlife as symbols of a passing frontier and its concomitant elements of American whiteness and manliness. Westerners fell into conservation almost by accident, led by opportunities for profit.[26] The particularities of ABS fund-raising revealed how Hornaday and other conservationists worked to cultivate a productivist ethos in order to blend in with the western political economy. They demonstrate how conservation, both as a national movement and as a matter of federal policy, emerged from local colonial efforts to dissociate predation from production. The ABS accomplished this feat by framing its own labor as conservationists in opposition to the work of predators—both animal and human. In so doing, ABS members simultaneously established themselves with credentials as western producers while using the logic of predation to build support for their conservation efforts.

Like Theodore Roosevelt, Hornaday sought to capitalize on representations of proletarian western labor while performing little of it himself. Although he held a reputation for "farmyard crudeness" with Roosevelt and other members of the New York elite, in his middle age Hornaday cultivated a white-collar image rooted in antiunionism and the affectations of northeastern old money.[27] As director of the New York Zoological Society for thirty years, Hornaday spent his days in the Bronx fund-raising and promoting the zoo, not feeding the park's animals or sweeping up the peanut shells left by visitors, a "curse," he sulked, that "fearfully disfigured miles of walks and lawn borders."[28] Hornaday's conception of productive labor seems to have corresponded closely with his own daily activities. Raising and allocating money for conservation was Hornaday's profession and the measures by which he judged his own and others' productive work.

Hornaday documented his fund-raising efforts well in the ABS's official history of the founding of the National Bison Range. Although Congress provided funds for the purchase of the Flatheads' surplus unallotted land, the ABS was responsible for raising money to purchase bison to stock the park. From his office in the Bronx, Hornaday coordinated a national subscription campaign to drum up $10,000 for the fund, a fund-raising effort that had succeeded by the end of 1909.

In the course of the campaign, however, Hornaday encountered resistance from western businesspeople, and in the end, the bulk of the bison subscriptions came from New York and Massachusetts, the western states failing to "contribute becomingly" to bison conservation.[29] "The men of the West are all right when it comes to commercial development," complained Hornaday, "but in a matter requiring as broad citizenship as the founding of a National Bison Herd . . . it is the men of the East, who as usual, bear the burden and heat of the day!"[30] The society's secretary, Frederic Kennard, politely suggested that Hornaday tone down his criticisms of western philanthropy in the society's official history. But with characteristic bombast, Hornaday simply responded: "When a man starts in to write history, he is expected to state the facts."[31] Hornaday undoubtedly felt frustrated and betrayed by the lukewarm reception his conservation plans garnered in the northern plains states where the buffalo once roamed.

From Minnesota to Montana, businesses balked at Hornaday's plans. Having failed, as yet, to make the proper impression as a productive ally of western business interests, he attempted to shame skeptical donors into making contributions. His efforts in Minneapolis and St. Paul offer a good example of Hornaday's strategy. With a hint of Manhattan pretension, Hornaday wrote the Minneapolis Commercial Club asking for $250, the contract price of one bison from the Conrad herd. "If there are not at least 250 persons in Minneapolis who are willing to subscribe $1.00 each . . . then I am totally mistaken in my estimate of your people, and I will be willing to state publicly that I over-estimated Minneapolis!"[32] Hornaday's solicitation earned mention in the city newspaper but little in the way of funds.[33] Looking across the Mississippi, he approached St. Paul with similar ineffectualness. Offering a brief and inaccurate history of St. Paul as an entrepôt in the fur trade, Hornaday's letter to city officials suggested the "big furhouses should subscribe handsomely; for goodness knows, they made good profits out of the extermination of the Bison."[34] It was a vain and futile attempt to seduce donors by implicating the "furhouses" in the predatory destruction of America's bison.

Unsurprisingly, no one from Minneapolis or St. Paul donated to the ABS subscription fund, with the significant exceptions of James J. Hill and Howard Elliot, owner and chairman, respectively, of the Great Northern and Northern Pacific Railroads. Taking a sweeter tone with these two business magnates, Hornaday emphasized how donations to the Bison Range would enhance the two men's philanthropic reputations. But even so, Hornaday incorrectly estimated their levels of interest in his bison conservation project. Tied closely with St. Paul's commercial community, Elliot acquiesced to Hornaday only after the fund-raiser berated him and his city for their lack of charity. Elliot reluctantly provided an anonymous donation of $1,000, a full 10 percent of Hornaday's fund.[35] His competitor, James J. Hill, already owned his own personal herd of bison. Hill had little interest in supporting a bison herd adjacent to the Northern Pacific's rail line but nowhere near Hill's line through Glacier. In the fall of 1908, however, Hill agreed to donate seven of his bison to the project. None of these bison ever made it to the refuge, however.

The failure of Hill's donation revealed the tensions that racked bison conservation outside of eastern urban centers. Hornaday, Roosevelt, and other early popularizers of bison had situated the animal as a symbol of frontier manhood for an eastern elite afflicted by neurasthenia, a reminder of their pioneer ancestors' supposed colonial virility. However, on the rural plains and prairies, local farmers and ranchers predominantly viewed the bison herds of Hill and others as affectations of wealth and pretension. Hill acquired his bison in the late 1890s from U.S. senator Richard Pettigrew of South Dakota. The animals arrived in St. Paul safely by rail, but Hill's livery staff, inexperienced with bison, were unable to herd them to his estate in North Oaks. Local western knowledge, as well as affect, were essential for managing the animals. Hill asked Pettigrew for advice; Pettigrew recommended Thomas Hardwick, an old Whoop-Up Country wolfer approaching seventy years of age and notorious for his role in the Cypress Hills massacre, for the job. Hardwick traveled to St. Paul at Hill's expense, telling the railroad baron that he could herd the bison by himself with only "a good horse and saddle." Mounted on Hill's private riding horse, in just five minutes the old-timer drove Hill's bison the remaining mile to their enclosure.[36]

The inexperienced staff at Hill's farm could have used Hardwick's full-time assistance, since it was difficult to keep the bison at home on the estate. On December 14, 1910, one of Hill's bison bulls escaped its enclosure and embarked on a forty-eight-hour tour of rural Anoka County, invariably described by local newspapers as a "rampage." The sensational incident revealed the seams of rural resentment toward Hill's genteel menagerie. In violation of federal bison-hunting restrictions, local farmers and police began stalking the bull, nicknamed Bad Bob by the county press. "Not a farmer young or old remained at his work," reported the *Tribune*, "but got out with the nearest deadly weapon . . . everything from a pea-blower to a Gatling gun."[37] The owners of National Automobile in Minneapolis donated the use of their new forty-horsepower race car, the Arrow, for the chase. After tracking the animal through Anoka's fields and woodlands for an astonishing two days, over fifty hunters were present for the animal's expiration in a farmer's ditch on December 17, when Bad Bob bled to death, crippled by several misplaced rounds of rifle fire. Hill's foreman requested and paid for the return of Bob's carcass to North Oaks.[38]

Reluctant to part with his prestigious animals, Hill delayed shipping his pledged bison to the National Bison Range for over two years. Hornaday wrote Hill several times asking for the status of the bison shipment, only to hear excuses about the expense and difficulty of shipping the live animals to Montana.[39] On the eve of Bad Bob's escape, Hill's son apparently had several of the bison slaughtered and fed to guests at a party. "It is a great pity that any female buffaloes should be slaughtered," lamented Hornaday.[40] In the end, Hornaday asked Hill to transform his pledge into a cash donation, but Hill refused. By the close of 1910 the ABS had warmed neither the hearts nor the pocketbooks of western businesspeople and philanthropists. Learning from Hornaday's mistakes, however, other members of the Bison Society would perfect its narratives of predation and production over the following decade.

West Meets East: Writing the Narrative of Productivity and Conservation

Throughout Montana, local concepts of predation and production affected the central ideals of bison conservation and their effective implementation. The Bison Society faced similar challenges mobilizing funds in Montana, although Hornaday and other members eventually achieved a great level of success utilizing their personal connections to represent themselves and their work as economically productive.

Western American identity was firmly yoked to myths of frontier conquest and pioneer spirit during the early twentieth century.[41] For Hornaday and Bison Society members, L. A. Huffman, a Montana oil speculator, former state legislator, and prominent Montana photographer, particularly exemplified these connections. Described in the *Montana Stock Grower's Journal* as "a man who is right abreast of the flood tide of twentieth century progress," Huffman, as an artist and businessman, also capitalized on Old West sentimentality and his reputation as a Montana pioneer.[42] Huffman offered a hybrid figure for the Bison Society to resolve the tensions between conservation sentimentality and western progressive colonialism. From the late 1870s through the 1920s, his photographs captured the Montana plains' transformation from a wild grassland to a bounded landscape of farms, ranches, reservations, and oil pumps. Serving as the American Bison Society's informal Montana liaison, he also led its members in several failed investments in the Cat Creek oil region east of Lewistown, Montana. Huffman's activities offer a provocative window into the relationship between business and the conservation movement in the early twentieth-century American West.

Huffman and his colleagues reconciled conservation and development by appealing to an ethos of western manliness founded on labor and its relationship to the earth and upon which hung notions of Anglo-Saxon racial dominance, as well as productivity and progress. Huffman and other white men and women of eastern Montana clung to a perception of themselves as hardworking people who labored to extract the resources of the northwestern plains—whether animals, metals, or fuels—in order to sustain, enrich, and grow their western communities. For eastern men like Hornaday, celebrating and preserving the symbols of this so-called pioneer heritage in the face of the industrial transformations racking the United States held importance. Huffman's photographs, primarily pictures of pastoral work, helped popularize the mythology of manly frontier labor. His unique images of bison hunting and his later images of ranch work featured white male laborers as brawny protagonists of independence and as heroic tamers of wilderness and animality—the latter a key component of colonial notions of predatory behavior. By contrast, Huffman's only published image of a non-Anglo ranch worker is of a man called Mex John working in a domestic role as a cook.[43] Like the work of Charlie Russell and Frederic Remington, Huffman's triumphal images of white manhood resonated widely across the United States and Canada, and his work achieved a fair measure of national celebrity. Theodore Roosevelt adorned the White House with six large Huffman prints (see figure 5). Cowboy songwriter Badger Clark lamented that "trails, guns, game, leather pants, sunburned noses and free-and-easy ways of living are all vanished or vanishing" but that he found solace in Huffman's images. And in the late 1920s, a teenager from South Carolina even wrote Huffman asking if he could move to Montana and work with him on the open range.[44] By that time, however, Huffman was an old man long out of the hunting and ranching business who made his living as a landlord in Miles City and Billings. Even so, Huffman retained his persona as a Wild West legend until his death in 1931.

With this high profile as a western artist, Huffman's connection with the ABS benefited the organization. Already close friends with William Hornaday from the former's hunting trips to Montana in the 1880s, Huffman's inclusion in this appendage of the eastern establishment provided the ABS with a much-needed local presence in Montana. And Huffman's photographic work also created a visual image of the cultural narrative that linked conservation with productivity and progress. With regional politics dependent on stock growers' associations, irrigated alfalfa farms, and railroads, wildlife conservation programs had more trouble finding supporters on the Montana plains than in the mountainous western third of the state. Huffman's reputation as a Montana pioneer provided a critical link between Hornaday's Wall Street–based alliance of conservationists and

5. *Five Minutes' Work*. Photo by L. A. Huffman. Courtesy of the Huffman Studio Collection, McCracken Research Library, Buffalo Bill Center of the West, Cody, Wyoming. P.100.3112.

local Montana cattlemen. Moreover, Huffman's pioneer identity provided a model for the ABS's mostly younger eastern members. These would-be prodigal sons and cowboys of the postfrontier generation admired Huffman's western authenticity and reveled in his rare visits to New York City. Introducing Huffman to their families and business colleagues whenever he was in town, these ABS members also referred Huffman's work to magazines and publishers and ordered more prints of his photographs.[45] Invited through one ABS member as a guest to a dinner honoring Theodore Roosevelt, Huffman dined on French haute cuisine and conversed with the Fairchilds, Dominicks, and other wealthy easterners eager to offset their masculine insecurities through an association with Huffman's western gruffness.[46]

Huffman also brought Bison Society members directly into Montana business affairs by selling them investment shares in start-up oil companies. Serving as the Bison Society's de facto Montana consulate, he helped members chart the regional terrain of commercial politics. Investing his modest fortune in the Montana Mutual Oil Syndicate and the Black Panther Oil Company, Huffman bought large blocks of shares and resold them to many of his colleagues in the East whom he knew through his involvement with the American Bison Society. In particular, Edmund Seymour—president of the American Bison Society following Hornaday's retirement—and George Roberts, both successful Wall Street bond traders, invested thousands of dollars in these ventures at Huffman's direction.[47] Located in the Cat Creek oil field, one of the most profitable oil districts in Montana, both the Black Panther and Montana Mutual ventures failed miserably. The Black Panther Oil Company was a particular source of aggravation for Huffman. Turned on to the brand-new, Philadelphia-headquartered business by his son-in-law, the Montana congressman W. T. Felton, Huffman sold large blocks of the company's shares to his local and eastern friends. Before the company dug any wells in Cat Creek, however, the federal government imprisoned several of its officers for fraud and embezzlement.[48] Huffman's investment in the Montana Mutual Oil Syndicate was a more private, local affair, but it ended in disaster too when the crude pulled from its wells in Cat Creek proved too impure to cover the piping and refining costs necessary to bring it to market.[49]

The Bison Society's participation in Montana's oil industry demonstrated that conservationists did not necessarily contradict western development; the two trajectories went forward hand in hand, based on their belief and faith in science and on the conviction that the manly labor of hunting and resource extraction—labor in the earth—was necessary for the continued vitality of the United States. Following the work of historian Frederick Jackson Turner, historical and social theory in the early twentieth century postulated that the United States derived its greatness from its frontier experience, that America was essentially a rejuvenation of European democratic institutions and culture forged in the savage wilderness of the American West. This frontier narrative offered a way of streamlining American business and politics, and the Bison Society sought to take full advantage of it.[50]

Perhaps more than any other eastern Bison Society benefactor, Ed Seymour perfected these western affectations and provides another example of the embodied link between western ideas of productivity and eastern ideas of conservation. Born into one of the Hudson River valley's wealthiest families, Seymour first met Huffman through Hornaday several years before the oil fiasco, when he had been searching out a .45-120 buffalo rifle for his gun collection.[51] Eager to collect relics of the nation's western heritage, he was also anxious about his own past history in the West. As a young graduate of Williams College in the 1870s, Seymour purchased a herd of five thousand cattle in Texas that he eventually moved to northern Wyoming. He quit the cattle business following the bust of the mid-1880s and used his remaining capital to open a bank in Tacoma, Washington. After marrying in 1891, he moved back to New York and began trading bonds.[52]

In his correspondence with Huffman and other western contacts, Seymour emphasized his western credentials. In courting one potential benefactor from San Angelo, Texas, Seymour explained his personal history in the Texas cattle business. Although he admitted moving his herd to the northern plains, he lauded Texas as "a wonderful state . . . doing some very good things in the preservation of game."[53] Unlike Hornaday, Seymour was careful to compliment potential western allies, encouraging them to consider themselves as stakeholders in regional wildlife preservation through their shared commercial interests in cattle, oil, and land development. Seymour himself represented these shared interests well. He served simultaneously, for instance, as president of the ABS and as the chairman of a national livestock corporation. In forging alliances in Montana, Seymour recognized the importance of valuing local perspectives, but he also made it clear that association with the Bison Society would increase the political clout of Montana organizations. In a letter to one Missoula business owner, he explained that, while "you can function usefully in the West . . . you never will have influence east of the Missouri River to amount to much." Joining the Bison Society, urged Seymour, would broaden the scope of Montanans' potential in national politics. And doing so would come at small risk of compromising western ideals, because conservation was not and need not be antagonistic to local goals of productivity. After all, "I am a 'Wyoming Pioneer,'" reassured Seymour, "and I might be classed as a Montana Pioneer, etc."[54] Using his western background to demonstrate his commitment to the region's logic of production and predation, Seymour instilled trust in a regional elite initially suspicious of the ABS's intentions as a conservation organization.

But performing western business culture was not, on its own, sufficient to establish conservation's compatibility with Montana commerce. The Bison Society and other conservation groups still needed to demonstrate their willingness to serve as good neighbors to the cattle industry and other regional enterprises. Managing wildlife refuges as ranches was one way they accomplished this task. This primarily meant a commitment to killing wolves and coyotes, but it also meant harassing the colonized, such as Indians and Basques, as vicious predators. Learning from the dissociations of predation from production perfected by the livestock industry, conservationists further positioned their work as an acceptable labor of production through their attempts to designate, control, and remove predators.

Unnaturalized Immigrants: Racializing the Narrative of Predation

The eugenicist Madison Grant, author of *The Passing of the Great Race*, was one of the Bison Society's most prominent members, as well as an officeholder on the New York Zoological Society's board of directors. Grant was also one of the organization's most passionate defenders of western ungulates, especially antelope, and one of its most vociferous opponents of hunters. In 1919, while touring eastern Oregon as part of an exploratory trip for an antelope refuge with Secretary of the Interior E. W. Nelson and U.S. senator Harry Lane, Grant offered a $200 bounty for the conviction of anyone caught killing the animal—presumably Basque sheepherders.[55] Even by the standards of the uneconomical bounties placed on wolves during the era, this was a remarkable sum. Ironically, the majority of Oregon's poached antelope were killed for use as bait animals by bounty-hunting wolfers. But if Grant's bounty therefore exposed the subtleties of predation as an intellectual concept, it steamrolled any ambiguities with the weight of its payout. Using their prestige as scientific experts, Grant and other members of the ABS largely shaped western perceptions of predation as a biological category relevant to human and nonhuman populations during the 1910s and 1920s. At the heart of these conceptions was an understanding that predatory behavior was an immutable characteristic of particular human races and animal species, a concept rooted in the taxonomic, natural history paradigm of the late nineteenth century.[56] In the western and eventually national contexts, Basques, Indians, and wolves stood in as these natural predators—beings whose racialized bodies marked them as inherently violent, vicious, and wasteful toward wildlife (see figure 6).

With an eye toward eugenics, Grant, Hornaday, and other Bison Society members conflated the biological management of human and nonhuman populations.[57] Hornaday's *Wild Animal Interviews*, a collection of short animal allegories, offers an interesting manifestation of these ideologies of animals and race.[58] A strangely serious attempt to collect ethnographic testimony from different wild animals, *Wild Animal Interviews* is noteworthy for the ways in which Hornaday invested his animal subjects with racial characteristics. The bison and "star-spangled Antelope" interviewed by Hornaday seem convincingly civilized and, like the "Anglo-Saxon race," in constant danger of destruction at the hands of other savage species. Hornaday's interview with a gray wolf anthropomorphized the predator as a vicious savage, "the meanest and cruelest animal of all North America, and if I could do it I would exterminate all of you but one pair."[59] Hornaday's genocidal rhetoric resonated with western artist Frederic Remington's

6. Work of the unnaturalized Basque sheepherder. From *The Annual Report of the American Bison Society*, 1922, 48, 63, box 1, folder 23, L. A. Huffman Studio Collection, BBHC.

famous diatribe against so-called non-Americans: "Jews, Injuns, Chinamen, Italians, Huns—the rubbish of the Earth I hate—I've got some Winchesters and when the massacring begins, I can get my share of 'em, and what's more, I will."[60] Artists, conservationists, historians, and philosophers of the American West, as well as eugenicists and racial theorists, all contributed to these shared discussions of the scientific and aesthetic merits of whiteness based on an implicit understanding of predation as a savage characteristic.

This discourse of predation and race also inverted Anglo-American ideologies of Native and immigrant during the early twentieth century. Although Indians had been vilified as primitive and savage in earlier centuries by European Americans eager to maintain their connections with their mother countries, "native American" acquired a new nationalist meaning amid the new immigration of the Progressive Era. The new nativism originated during precisely the same decades when the United States and Canada intensified their bitter programs of conquest, colonization, and forced assimilation, which killed and dislocated thousands of North America's indigenous people.[61] Encapsulating the spirit of both Grant's and Hornaday's conceits of Anglo-Saxon dominance, by the Progressive Era "native American" had come to epitomize the second-, third-, and fourth-generation white American elite, often with British or Dutch surnames—the Roosevelts, Seymours, and Remingtons who squared off against the huddled masses in order to deal with their own class affliction, the racial and gender insecurities accompanying their rise to affluence.

Embracing the Indian, then, at least a fictitious one, not only offered a possible and popular way to reconcile these two meanings of "native" but also provided a mechanism for white elites to reclaim and celebrate their masculinity. Taking a trip to Oklahoma to kill predators on the Wichita Mountain Wildlife Refuge, President Roosevelt "acquired in the Indian country a complexion that would do credit to an Apache warrior," the *Washington Post* approvingly reported. "For the next four or five weeks he will make life miserable for members of the cat and bear family that happen to come his way."[62] These indigenous Progressive Era metaphors created a new category in nativist America's public memory and in its national self-identity.

By the 1920s the seeming success of Indian assimilation and citizenship also provided a counterpoint to problematic populations of "unnaturalizable" immigrants to the western states. This sentiment had been building in the Northeast for a number of years. In 1908, for instance, George Bird Grinnell wrote Huffman, the Montana photographer, looking for some photographs of "western scenes" for his magazine *Forest and Stream*. "The picture of a lot of dead animals hanging up, or with the slayers standing by them, do not appeal to us very strongly," wrote Grinnell, presumably familiar with Huffman's work on bison hunts. "Neither do we care for pictures of Indians," he continued, "unless they are doing something, and Indians wearing the war bonnet are barred."[63] It is unclear what "something" Grinnell wanted the Indians to do, but his prohibition of the war bonnet precluded the typical Remington and Charlie Russell scenes. Hunting for western images, but not those of savage Indians or slaughtered wildlife, Grinnell's editorial taste seems to indicate a broader Progressive Era admittance of notions of "civilized" or assimilated Indians into the mainstream.

These anxieties over nativity, predation, and savagery interacted with legacies of western conquest, ideologies of eugenics and conservation, and notions of virile white masculinity to create a complex xenophobia of immigrant Basque sheepherders. First arriving around 1900, Basque immigrants, many of whom worked with sheep in the high plains and mountains of the Pyrenean Basque country, found work in the interior northwestern United States as sheepherders.[64] Competing with struggling Anglo-Saxon cattlemen over dwindling range resources, the Basques, claimed cattlemen, upset the state's tenuously balanced grassland economy. More significantly, however, Basques provided a foil for the construction of a white nativism across Montana and the northern plains. Drawing on racial theories that classified Basques as non-European, Grant and the ABS singled out Basque immigrants as environmentally destructive, the keepers of animals that cattlemen ironically referred to as "locusts." While sheep typically graze grasses more closely than cattle, beef cattle greatly outnumbered sheep on the northern plains and took a far more drastic toll on the western rangelands. The 1924 Immigration Act set a quota on immigration from Spain, partially in response to these western grumbles over the Basque sheep industry.

In 1908 Huffman and Hornaday coauthored a photographic essay in *Scribner's* that offers a glimpse of the connections between bison conservation and suspicions of the sheep industry. Titled "Diversions in Picturesque Game-Lands," the article provided an account of a hunting trip that Hornaday and Huffman took the previous year, a trip to "energize" the two aging men—Hornaday weighted down by the "effête" surroundings of his office in New York, and Huffman apparently freed by the recent engagements of his two daughters.[65] Describing the environmental destruction wreaked by sheep in the Missouri River badlands north of Miles City, Hornaday lamented, "The awful sheep-herds have gone over it, like swarms of hungry locusts, and now the earth looks scalped and bald, and lifeless. To-day it is almost as barren of cattle as of buffaloes, and it will be years in recovering from the fatal passage of the sheep."[66] On their hunting trip, Huffman and Hornaday also found some dinosaur bones in a cutbank along a Missouri tributary. Sending the specimens to Henry Fairfield Osborn, Hornaday's paleontologist colleague at the New York Zoological Park, the bones led to the discovery of *Tyrannosaurus rex*, the northern plains' most apocryphal predator.[67]

H. F. Osborn was a friend of Madison Grant's and an advocate of Grant's racial taxonomy and eugenic history of America. In the preface to Grant's *Passing of the Great Race*, a call to address the problems of white "race-suicide" precipitated by the destruction of World War I and decreasing white birthrates, Osborn concluded that the "conservation of that race which has given us the true spirit of Americanism is not a matter either of racial pride or racial prejudice; it is a matter of love of country, of a true sentiment which is based upon knowledge and the lessons of history rather than upon the sentimentalism which is fostered by ignorance."[68] Both Osborn and Grant agreed that this patriotic racial character came from the "Anglo-Saxon pioneer type," an artificial evolution of man spurred by a Turneresque frontier conquest. Separating European whiteness into three races, Nordic, Teutonic, and Alpine, Grant's taxonomy ranked Nordic as the purest disposition of pioneering impulse, the basis of American racial character, and the natural human aristocracy. Basques, on the other hand, fit into none of these three categories. Due to their unique language, Grant theorized the Basques as an ancient "An-aryan" race migrated from Asia, some kind of hopelessly antiwhite presence in Europe.[69] Grant's analysis of Basque racial origin, written in 1916, established a seemingly scientific basis on which to discriminate against Basque claims to legitimacy as white co-owners of the American West, accelerating calls for their removal.

In its 1922 report, the ABS opined that "if the Basque sheepmen are allowed to occupy the country, it will be but a short time before the antelope are killed or driven off and the nests of the sage grouse trampled out by the sheep, vegetation destroyed, and the country made a barren waste."[70] The report also mentioned "60 antelope wiped out last winter by wolves and Indian dogs." "Stockmen pay taxes on land and cattle," the report continued. "They are opposed to the sheepmen and their methods, as he is neither a settler, desirable person or of any benefit to the community, pays no taxes, ruins the land and then passes on. . . . When the grazing is destroyed by the sheep, it is, and will remain, a barren desert of loose stones, whereas cattle have and do graze on it year after year without injuring it." Clearly, these tirades against Basque sheepherders represented the anxieties of white cattlemen toward immigrant sheepherders. But the metaphorical connections between Basques and wolves also served to naturalize Basque racial characteristics as savage and cruel.

Frank Van Nuys's *Americanizing the West*, a history of immigrant naturalization programs throughout the western states from 1890 to 1930, uses a hydrological metaphor to explain the history of the West's relationship with immigrants. "Like engineers harnessing a wild river's flow behind an elaborate system of dams," he suggests, "the Americanization movement had attempted to control the chaotic, diverse, and mobile immigrants of the West."[71] Van Nuys concludes that these efforts to Americanize immigrants—to naturalize them—as part of a contained human current failed in large part due to the "uncertain results" of Progressive Era immigration reforms.

But the Immigration Act of 1924, a system of quotas that relied on Grant's hierarchy of European whiteness to exclude non-Anglo-Saxon immigrants from southern and eastern Europe, was not just due to Basque failures to "naturalize" to the norms of American behavior. Instead, the obstacle to their

"Americanization" lay in the science of racial taxonomy and its imbrication within a logic of predation that cast immigrants such as Basques as unproductive species unworthy of inclusion in America's modern, scientifically managed landscape. For Basques, detested by conservationists as wolf-like predatory criminals, one obstacle to "naturalizing" as American citizens was their naturalization as predators. Racial ideologies bolstered by myths of white conquest and animal metaphors precluded the incorporation of Basques into the white conceptions of the western landscape.

Within a broader political ecology dominated by the livestock industry, wildlife conservationists had to make it clear that they would be good neighbors, and they achieved this mainly through persecution of predators. And in this way, the narratives of predation and production used by conservationists were shaped by the needs and goals of local westerners, who above all desired to maintain their own status as producers by continuing to designate and destroy predators. For instance, one woman who lived adjacent to Wainwright Buffalo Park during the 1910s, which she called the "coyote reserve," complained bitterly about the shooting restrictions within the park that had made it a seeming sanctuary for local predators. "The wolves have learned to come out and help them selves to the poultry and dash back in again," she explained. "With this park, I look upon the Dominion Government as a neighbor of mine who is not being neighborly."[72] In response to complaints like these, park managers and conservationists not only redoubled efforts to eliminate predatory animals from wildlife refuges but also extended their expertise to lay the blame for conservation difficulties on Basques and Indians—the local disenfranchised people who had been recast over time, both culturally and materially, as "predatory." This process of racializing predation was supported by the demands of integrating the eastern-based conservation of wildlife with the commercial prerogatives of regional western authority.

Hardening the Logic of Predation

"Advancing civilization" was a well-worn euphemism by 1930, when Stanley Young explained to Arthur Carhart why wolves would be annihilated. Just three years earlier, the two had collaborated on *The Last Stand of the*

Pack, a collection of stories celebrating the Bureau of Biological Survey's wolf hunters and their elusive quarry.[73] Recently promoted to the bureau's chief of Predatory Animal and Rodent Control (PARC), Young now faced a barrage of questions from Carhart, a former National Forest Service employee, an old friend, and a new advocate of wolves. Carhart's interrogation threw Young into a lucid counteroffensive, to the extent that copies of Young's response circulated throughout the bureau for the next thirty years, providing an official list of dogmatic answers to the difficult questions that scientists, conservationists, and the public asked about the agency's predator eradication programs.[74] Unable to support his position with empirical data—PARC, after all, waged "campaigns" rather than "studies"—Young used the familiar narrative of conquest, of "advancing civilization," to point out wolf defenders' central misconception: "You state, 'Isn't it a just consideration that the cats and wolves and coyotes have a damn sight better basic right to live in the hills and have the use of that part of the world as their own than the domestic stock of the stockmen?' Do you say that also with respect to the American Indian? Your reasoning in this connection would argue that the Indian should have been allowed to 'stay put,' but advancing civilization has dictated otherwise, and the same holds true for predators."[75] Young summoned the full force of frontier history's mythic inevitability to justify PARC's mission, a wave of civilization regretfully purging the plains of its unproductive predators.

Young was not unique in his outlook. From the beginning of the Progressive Era, American and Canadian scientists, activists, and bureaucrats utilized the discourses and practices of conquest to express the intangible values of wildlife conservation programs. In the Montana-Alberta borderland, bison and antelope had stood in as positive symbols of these national narratives, while wolves presented regretful figures of failed evolution under the forces of civilization. For organizations like the ABS, which actively created bison and antelope refuges throughout the Canadian-American West, wildlife conservation offered a way to atone for the sins of Anglo-American colonization while redoubling its political consequences. Saving a species nearly wiped out by commercial exploitation, Hornaday, for instance, saw in his work an effective battle to preserve living animal specimens of

North America's precolonial past as a way to demonstrate capitalism's ultimate governance by philanthropic principles. But driven by much the same motivation as their "salvage anthropology" contemporaries, the ABS sought wildlife for its representative power, as objects of colonial domination, rather than fighting to let the "uncivilized" live on their own terms. For the ABS and similar organizations, working to eradicate wolves was an easy decision; as predators of bison, wolves, like the Indians whose cultures anthropology fetishized but whose presence it disdained, held no physical place in the borderland's postconquest order.

The North American scientific mode of natural history also posed limits to the acceptance of wolves as valuable creatures. Dominated by descriptive methods that classified animal behaviors based on immutable characteristics, research in the field was slow to accept the dynamism of nonhuman behavior and remained unfocused on explicating the mechanics of organismal relationships. Under the influence of this philosophy, most zoologists and even scientists in the emerging field of ecology were content to attribute to wolves a set of static behaviors, namely, an insatiable taste for killing. Although Young, for instance, wished that "the wolf," always in the singular, "would change its ways just a little so that the hand of man would not be raised so constantly against this predator," wolves' failure to do so was a foregone conclusion: "This animal is one-hundred-percent criminal, killing for sheer blood lust, more often killing to satisfy his lust than to satisfy a natural and reasonable hunger."[76] Even the American Society of Mammalogists, one of PARC's earliest and harshest critics, admitted that wolves "were truly killers."[77] This dominant understanding postponed serious research into wolf behaviors until the second half of the twentieth century. Meanwhile, it hindered the scientific community's effectiveness as an advocate for wolf conservation.[78]

Usual narratives of wildlife conservation in the United States and Canada commonly locate the popular emergence of predator advocacy alongside Aldo Leopold's and Olaus Murie's epiphanies in the 1940s. Leopold's "land ethic" and Murie's abandonment of the Bureau of Biological Survey—like Young, Murie was a former PARC district manager in the Rockies—supposedly signaled the emergence of a new environmental movement from within the professional cadres of American universities and public agencies, establishing a latent core of sympathetic experts for an ecologically minded public to tap into by the 1960s. The best of these historians have analyzed in detail the opposition of scientists and conservationists to predator control as early as the 1920s, citing the "great rift" between the bureau and the American Society of Mammalogists over predator eradication, a battle that laid the foundations for an environmental movement guided by research in animal ecology.[79] But in celebrating the early dissenters and the eventual fruit of their efforts, historians of environmentalism too often forget that in the interim, the bureau won. In 1931, after years of intense debate, PARC secured a ten-year congressional commitment to fund and expand its eradication work over millions of acres of public land. The reasons for the bill's passage remained incomprehensible to its early environmental opposition, which was crippled by an unawareness of predator control's deep imbrication within the colonial logic of conquest and production, a logic that structured the opposition's politics as well.[80]

Under siege by the American Society of Mammalogists, an organization in which he himself was a member, Young faced his colleagues' concerns that PARC's lethal control methods not only hindered the scientific study of carnivorous mammals but also upset nature's ecological balance. Pointing out the absurdity of this latter objection in a world already mauled by industrial agriculture, Young remarked to Carhart: "In this day and age of man, wild life is where you find it." Recognizing, before his time, that wildness was part cultural artifice, Young placed his faith in an even more normative bastion of scientific logic: the difference between consumer and producer, between predator and prey. In seeking to establish the limits of "natural and reasonable hunger," Young dedicated his intellectual life to sorting out the moral boundaries of human and animal exploitation through the concept of predation.[81]

Stanley Young was not alone in his contemplations of nature and social order among modern American scientists. His interest in unraveling the moral bases of hunger and satiation reflected a powerful American concept of predation that was rooted in the colonial experiences of the late nineteenth-century West and leavened by a liberal-capitalist consensus in the early twentieth century. The 1931 passage of the Animal Damage Control Act operated on the principle that wolves, like other true predators, were voracious killers wholly unproductive according to the logic of modern capitalism. This principle was informed by a natural history paradigm that categorized the wolf as an animal with immutably violent behaviors. Just as aspects of the 1924 Immigration Exclusion Act hinged on the naturalization of Basques as predators, the Animal Damage Control Act established wolves, along with coyotes, mountain lions, and bears, as predators in need of exclusion from the western landscape.

Conclusion

Well into the 1900s, wildlife conservationists and policy makers were the heirs of a logic of predation that evolved amid the settlement and colonization of the northern borderlands. This logic, more than the broad cultural milieus of frontier and empire, served to justify and legitimate the creation of wildlife preserves in Montana and Alberta during a period of major land contestations. In fact, as chapters 3 and 4 have demonstrated, the logic of predation itself gathered momentum during the early years of these controversies—particularly within processes of reservation confinement and allotment—because of the ways that predatory logic separated land users through perceptions of productivity. Collapsing predation with Anglo-American concepts of plunder was one method by which white cattlemen represented themselves as producers and, as such, the rightful occupiers of the northern rangeland. This same logic also provided conservationists the discursive means to justify their incursions on stockmen's so-called productive lands for the sake of rehabilitating wildlife. For both eastern conservationists and western commercial interests, the narrative of production and predation had proved their utility in supporting the colonization of western lands and marginalizing competing peoples and animals.

Despite their colonial affinities, however, tensions between wildlife conservationists and the stock industry were not easily surmounted. Bison conservation was driven by an uneasy and uneven cooperation between eastern sentimentalism and western business interests. As Andrew Isenberg has noted, bridging this divide required an accommodation where conservationists purchased bison herds from enterprising regional stock owners and confined their activities to isolated refuges that did not threaten the livestock industry's broader political ecology. The profit motive, or "the returns of the bison," as Isenberg phrased it, connected the divergent interests of stock growers and conservationists, allowing the possibility of bison preservation. Taking a different approach, however, this chapter has argued that the origins of this compromise exist in the logic of predation, revealing how William Hornaday and the American Bison Society relied on framing their agenda as productive in order to attract western support for their conservation efforts.

Beyond simply creating the political possibility of conservation, this logic of predation also set limitations on conservationists' possible achievements. Jennifer Brower's recent history of the Wainwright Buffalo Park contends that its failures were due primarily to a lack of professional expertise in ecological management. "As little was known about wildlife science until the 1930s," she explains, "the Parks Branch turned to sources knowledgeable in domestic animal management and relied heavily on the Department of Agriculture for advice on managing the bison."[82] Although Brower may be correct in emphasizing the Canadian Parks Branch's scientific ignorance, this chapter suggests instead that the bureaucratic drift of wildlife conservation was already overdetermined by its proximity to predator-producer idioms of the livestock industry that had evolved throughout the region's local history. Conservationists applied agricultural solutions because the logic of predation foreclosed other possibilities, such as the reintroduction of bison to private rangelands.

Underpinning these specific horizons of wildlife conservation was a broader narrative of predation that ascribed animal extinction to the contingencies of overhunting rather than environmental transformations effected by colonization and habitat loss. As this chapter demonstrates, conservationists like the ABS created and utilized this narrative not only to drum up "sentiment" for the preservation of wildlife but also to create a mythology of animal extinction that would play well throughout the cattle-crossed borderland, from Billings to Banff. William Hornaday himself authored the first history of bison extinction in 1889, attributing their disappearance to the work of vicious market hunters. Hornaday's history, like the six bison he shot and stuffed for a Smithsonian diorama, became emblematic of the American conservation movement and its legacy. Although excellent research by Ted Binnema, Dan Flores, and Andrew Isenberg has considerably revised the role of grassland ecology in the historical fluctuations of bison populations, the vast historiography of American bison is still dominated by lamentations of their wanton destruction. As recent debates over the "sustainability" of historical bison hunting by indigenous people reveals, the literature still mainly deals with assessing blame to individual agents rather than understanding the broader roles played by capitalism and colonization in the animal's histories of near destruction and slow recovery.

As the ex-officio president of the American Bison Society in 1913, Hornaday decided that the organization had fulfilled its mission of saving the bison. "The future of the American Bison is now secure," he voiced privately, "at least so far as non-extermination is concerned."[83] In celebration of the conservation successes, the society hosted a dinner at Delmonico's. Although bison was not on the menu, they ate a five-course meal comprised of chicken, halibut, squab, and, as the entrée, a roast fillet of beef.[84] Eating bison once again would not be far off. In 1925 the Northern Pacific reintroduced the animal to its dining cars, as Isenberg has observed in his history of bison conservation. Serving culled animals from the National Bison Range, located just on the north side of the tracks, the railroad's roasts and steaks "were as savory and tender as the Indians and pioneer hunters ever tasted."[85] Wildlife conservation was about more than just limiting the kill; it was about establishing who could do the killing.

Conclusion

> At times we will have to howl with the wolves, and that means that we will have to forget some things that are familiar.
> —Hans Peter Duerr, *Dreamtime*

It takes between ten and twenty pounds of grass each day to feed an elk during winter. The animals migrate down from alpine meadows to winter rangelands, kicking divots through the layered crusts of ice and snow, lowering their warm mouths to the earth to graze on frozen plants below. As winter temperatures plummet, the elk grow leaner with each day's caloric losses. The more fat they burn, the more grass they need to eat. Life is difficult for elk that start winter wounded and exhausted from fighting and mating during the fall rut and even more so for their calves and yearlings. Weakened by cold and starvation, they are easy prey for wolves, who follow the herds down from the mountains and cull the year's crop of runts and stragglers.

For the wolves that returned to Yellowstone National Park in the winter of 1995 after a six-decade absence, hunting was even easier. They found elk that had no fear, elk who did not know they were prey. In the technical nomenclature of wildlife biologists, Yellowstone elk were "predator-naïve."[1] Not only did they lack predator-avoidance behaviors to structure their grazing and herding patterns, their physical bodies were underdeveloped and less able to escape attack. Wolves surrounded and slaughtered the healthy in addition to the lame, startling wildlife scientists and policy makers who had underestimated the numbers of elk that wolves would kill. Like other animals, when presented with an overabundance of food, wolves killed many more animals than they could eat, a phenomenon that scientists called "surplus killing." Park biologists predicted that these incidents would abate as elk grew familiar with the danger of wolves in Yellowstone, but as wolves expanded their territory, the surplus killings spread. In one incident, a pair of wolves killed 120 sheep in a summer pasture near the Blacktail Mountains of southwestern Montana, consuming portions of only two carcasses. Local livestock growers condemned the episode as a vicious and wasteful slaughter.[2] For sympathetic observers of wolves, these events were troubling and unexpected. The habits of wolves were more contingent, historical, and cultural than they were instinctual. Predator and prey were not a priori biological certainties but social categories worked out between wolves and other animals in the field.

We too are animals, and our lives are reliant on the flesh and blood of other creatures. As this book has demonstrated, however, the history of wolf eradication and the corresponding history of the colonial separation of predators from producers have accelerated our disavowals of this reality. Between the 1860s and the 1930s the conquest of the Northern Rockies frontcountry transformed the concept of predation from an indigenous understanding of a livelihood sustained by death into a colonial indictment of humans and animals whose labors did not conform to the new political ecology of livestock capitalism. The dissociation of predation from production first solidified during the 1880s, when public and private authorities instituted a series of bounty programs to encourage the extermination of wolves and other carnivores. Although largely ineffective for addressing the economic costs of livestock lost to wolves, the bounty system helped establish clearer lines between predation and production and in turn justified the cultural legitimacy of ecological practices that transformed the Northern Rockies into cattle country by the end of the nineteenth century. The subordination of indigenous land and labor occurred alongside this process. Colonialism relied on federal programs of assimilation that sought to purge Native people of their so-called predatory behaviors and to restructure their environmental practices to conform to new systems of wage labor and private landownership. From the early twentieth century onward, wildlife conservationists appropriated this logic of predation and production to establish themselves as stewards of animal production rather than predatory degeneration.

The outcome of this history is that we can only love wolves when we represent them as producers; we cannot love them for being predators. When wolves kill cattle and sheep, the state responds by shooting them from helicopters. When wolves howl to delighted nature-goers, officials salivate at a tourist economy of wolf watching that generates an estimated $23 million each year.[3] Wolves have long played starring roles in ironic celebrations of the West's legacies of conquest. For many individuals, past and present, wolves' location on the seeming edge of civilization has worked as a stand-in for their own feelings of social marginalization. Jody Emel, for instance, a self-described ecofeminist, writes that "wolves are symbols of resistance. These animals are metaphors for oppositional ways of thinking and feeling."[4] When people in the Northern Rockies look for wolves today (and seldom see them), they are often hoping to catch a glimpse of this recalcitrant past, expecting to hear some defiant lupine dialogue.

The problem is that wolves subvert the spaces we set aside for them. Ewen and Evelyn Cameron learned this lesson on their eastern Montana ranch more than a century ago when they took possession of two wolf pups from their hired wolfer, Maurice Barret. Rather than cash in the wolves for bounty payments, the Camerons decided to raise them as pets. But the wolves resisted the Camerons' ownership. In a manuscript rejected by *Scribner's*, Ewen recounted Evelyn's attempts to photograph women of neighboring ranches with the wolf pups posed on their laps. "Owing to their frolicsome ways," he wrote, "it was almost impossible to obtain satisfactory photographs of our charges since they declined to remain still for the fraction of a second, and could only be focused when asleep." Their resistance to being captured on slow-speed film was not merely due to their playful behavior, however. "When two months old," Cameron explained, "their innocent appearance constantly tempted lady visitors to try and caress them, but the ungracious reception accorded to these overtures soon repelled the most enthusiastic lover of animals."[5] About six months later, apparently fearing for Evelyn's safety, Ewen sold the pups to an amusement park on Coney Island, where the animals lived in captivity until burning to death in a fire (see figures 7 and 8).

7. Jetta Hamilton Grey with wolf pup. Photo by Evelyn Cameron. Courtesy of the Montana Historical Society Research Center Photograph Archives, Helena, Montana.

It has become intellectually fashionable to think of our relationships with nonhuman beings under the idiom of hybridity, but the history of our relationships with wolves throws the optimism of these theories into disarray. In one of Cameron's photographs, of a young wolf pup chained inside a box, who is being protected from whom? This is not a photograph of Donna Haraway's dog, Ms. Cayenne Pepper, the pedigreed Australian shepherd and agility training champion of twenty-first-century California "natureculture." Cameron's photographs depict a different and unwanted intimacy. They reveal how our companion-species relationships with wolves and other animals are deeply flawed. Haraway's brilliant scholarship is right, of course, to insist that we study the material entanglements that connect humans with the nonhuman world.[6] But we should be careful to avoid the preconception that wolves and other creatures want to be our friends. In order to truly liberate wolves, we need to let them live their own lives.

A decision to live is a decision to kill, and it can be difficult to love wolves

8. Janet Hogmire with coyote. Photo by Evelyn Cameron. Courtesy of the Montana Historical Society Research Center Photograph Archives, Helena, Montana.

when they make this choice, when they transgress the ethical simplifications we use to construct our sometimes naive visions of nature and justice. In 1983 the Walt Disney Company released its movie adaptation of Farley Mowat's popular novel *Never Cry Wolf*, a fictional account of a biologist's discovery that wolves subsist on mice rather than the frightened, slain, and shredded bodies of large mammals.[7] This narrative appealed to millions of viewers, including perhaps Timothy Treadwell, the amateur naturalist and filmmaker who followed in Mowat's footsteps. Treadwell traveled to Katmai National Park from southern California, living with grizzly bears in an attempt to document and publicize their gentleness.[8] In the fall of 2003, however, an *Ursus horribilis* from the interior killed and ate Treadwell and his partner. Although wolves are rarely "man-killers," like grizzly bears, real wolves are also different from the ones we discover on film. The peckish *Canis lupus* of Mowat's imagination turns out to be a more sanguine animal in the flesh.

What can we do to help wolves? For a start, we might reconsider dissociations between predation and production. The reproduction of life is not reducible to these biological mechanisms. In setting lines between predation and production, we have helped entrench problematic boundaries between human and animal and between civilization and barbarism, relics of conquest and colonization. As Alistair Graham noted in his controversial study of man-eating crocodiles, the usual language of conservation too often functions on an "inverted aggression and crippling sentiment" that limits our abilities to love other animals without casting them in allegories that justify our own uneven social relationships.[9] Humanity's modern emergence as a species of "supermarket carnivore," as David Petersen has put it, has undoubtedly relied on dissociations of predation from production that have been rooted in the historical development of colonialism and capitalism.[10] Rather than admitting the labor of killing, we have lived in luxurious alienation for far too long, producing animals for meat in secret and under conditions of horrific cruelty. Compassion for the lives that end in order to sustain our own can only come from acknowledging our own complicities in predator-prey relationships, not by purposefully forgetting the deaths inherent to our modes of production. Maybe by "living like wolves" we can conceive of a more just and equitable world.

Notes

INTRODUCTION

1. See Peter Koch to his uncle, January 21, 1870, in "Letters from the Musselshell," 322.
2. See Koch, *Splendid on a Large Scale*.
3. The historian William Cronon, writing about the industrial stockyards of nineteenth-century Chicago, observed that "the growing distance between the meat market and the animals in whose flesh it dealt . . . betokened a much deeper and subtler separation—the word 'alienation' is not too strong—from the act of killing and from nature itself" (*Nature's Metropolis*, 212). Meat, of course, does not grow in the soil, it is carved from the bodies of animals. But as Noelie Vialles argued in her ethnography of slaughterhouses, the humane imperative of modern western meat making has been to dissociate meat from its animal origins, to "vegetalize" meat in order to mystify the violence inherent to its production (*Animal to Edible*). For these reasons, a garden metaphor for the northern plains' transformation into industrial cattle country seems appropriate.
4. In his classic study of the Columbia River, Richard White noted that "we conflate energy and power, the natural and the cultural, in language, but they are equally mixed as social fact." For travelers negotiating the rapids at Celilo Falls, for instance, home to one of the oldest indigenous communities in North America, "the energy system of the Columbia determined where humans would portage, but human labor created the actual route of the portage, and human social relations determined its final social form and outcome" (White, *The Organic Machine*, 14). The same might be said in the case of energy transfers that take the form of predator-prey relationships and combine natural metabolisms with contingent

cultural processes. The source of value may be nonhuman, but its ecological flow is never automatic and requires constant political struggle both in the realm of environmental technologies and in venues of cultural representation.

5. This idea of an interspecies rivalry pitting wolves against people in a violent quest for predatory domination is a narrative commonly taken up by the vast literature on human-wolf conflict. Unfortunately, many environmental commentators writing about the history of wolf eradication have stopped short of assessing the particular social contexts in which these campaigns occurred, instead understanding human-wolf relationships as a timeless antagonism that coalesced somewhere in the deep past of human civilization. But the best work in this field has contextualized these conflicts alongside specific sets of cultural transformation, especially those wreaked by modern forces of industrialization and capitalism. See Coleman, *Vicious*; Jones, *Wolf Mountains*; Lopez, *Of Wolves and Men*; Rawson, *Changing Tracks*; Robinson, *Predatory Bureaucracy*; and Walker, *The Lost Wolves of Japan*. To a certain extent, this book follows the tradition of these works by focusing on how the privatization of landownership and the arrival of domestic livestock to the Northern Rockies put wolves in the crosshairs. But it also situates these conflicts into a broader social world that connected the whole enterprise of wolf eradication in the borderland to the dispossessions of the region's Native peoples. The most significant outcome of designating and killing predators was to justify the conquest of the Northern Rockies by legitimizing particular modes of economic and environmental exploitation as productive. The eradication of wolves was a critical element in this larger process of redefining work to conform within the constraints of colonial capitalism.

6. Bell, "Hunting Down Stock Killers," 290.

7. Language represents a long-standing and critical problem in the methodology and philosophy of animals studies in the humanities, and this book's theoretical approach owes a debt to these discussions. Scholars in the interdisciplinary field of animal studies have begun to question the place of language as the dividing line between humans and nonhumans. Shortly before his death, for instance, Jacques Derrida reinterpreted this problem by suggesting that it was not animals' inarticulateness but a trained human inability to listen outside of language that really mattered as a barrier to interspecies communication. Animal history "would not be a matter of 'giving speech back' to animals," as Derrida saw it, "but perhaps of acceding to a thinking . . . that thinks the absence of the name and of the word otherwise, as something other than a privation" ("The Animal That Therefore I Am," 395–96, 416). Likewise, the philosophies of poststructuralism and pragmatism have long suggested that words have no inherent meaning and that language is best understood as a tool for accomplishing an outcome, more

like a dog barking than a stable message of fact or truth (Rorty, *Contingency, Irony and Solidarity*). Rane Willerslev's brilliant ethnography of Yukaghir hunting practices illuminates these restrictions of language, even the limitations of the "verbal accounts" often cherished by anthropologists and oral historians, which, to the Yukaghir, represent "inferior ways of knowing compared to lived experience" (Willerslev, *Soul Hunters*, 159). In my view, it seems that imagining these experiential contexts of past human-animal interactions is a crucial element of analyzing documentary sources. It is analyzing practice rather than language alone that sets the methodological agenda for blending cultural and environmental history.

8. In this book, the terms *Blackfoot* and *Blackfeet* conform to their customary regional usage in the Northern Rockies. *Blackfeet* refers to those Blackfoot people connected to the Blackfeet Reservation of northern Montana. The Blackfeet today represent a substantial portion of the Pikuni, or Piegan, people, one of three tribes composing the Niitsitapi, or the "Blackfoot Confederacy" of the northwestern plains, an Algonquian society united by shared language, kinship, and custom and most commonly known in English as the Blackfoot—a reference to black moccasins. The Blackfeet Reservation was established by treaty in 1855 for Pikunis living on land claimed by the United States. In Canada, Treaty Seven, produced in 1877, established three other reservations for Blackfoot people: the Blood Reserve for the Kainah Tribe, the Blackfoot Reserve for the Siksika Tribe, and the Piegan Reserve for those Pikunis living north of the international boundary on land claimed by the Dominion of Canada. Because they are both acceptable and familiar phrases in English, I have decided to use the term *Blackfeet* to describe the southern Pikunis of the Blackfeet Reservation and *Blackfoot* to discuss the Niitsitapi more generally.

9. For a good overview of research related to these Blackfoot methods, see Barsh and Marlor, "Driving Bison."

10. Report of Inspector Thomas, "Inspection of Blackfeet agency and charges against Agent Allen," October 10, 1885 (5017), in United States Bureau of Indian Affairs, *Reports of the Inspection*, reel 2.

11. This observation may seem jarring in the wake of postcolonial scholarship that has sought to correct the scientific racism of past centuries by asserting the essential humanity of indigenous people. While western scholarship has recently begun to reconsider the question of what it means to be animal, "colonized peoples," writes Linda Tuhiwai Smith, "have been compelled to define what it means to be human because there is a deep understanding of what it has meant to be considered not fully human, to be savage" (*Decolonizing Methodologies*, 9, 11). Students of indigenous environmental practice should operate with this tension

in mind. But it is equally important to recognize the fact that the very distinction between human and animal is rooted in ancient Old World cosmologies and is not how most indigenous people of North America experienced or understood themselves in the precolonial era.

12. Several historians have explored this western mentality of improvement from a critical perspective. See Bantjes, *Improved Earth*; and Knobloch, *The Culture of Wilderness*.
13. David Rich Lewis has explored similar transformation in case studies of the Northern Utes, Hupas, and Tohono O'odhams, although his analysis focuses primarily on agriculture rather than hunting (*Neither Wolf nor Dog*). Chapters 3 and 4 of this book take up the environmental and labor historiography of Native Americans in more detail.
14. Bastien, *Blackfoot Ways*, 27.
15. My approach in this book to understanding predation and production as categories of valuing work has been influenced by a variety of theorists who have explored the idea of labor as an outcome of contested cultural representations of value rather than the idea of labor as an essential input to production. In other words, these scholars have focused on the messy cultural politics that delimit the boundaries of what kinds of work count as valuable or even count as work at all rather than analyzing and critiquing capitalist production from the standpoint of classical Marxian labor theories of value. These scholars have developed a perspective described by Diane Elson as a "value-theory of labor." Understanding work from this perspective offers a useful starting point to rethink colonialism in western North America as a process of disavowing indigenous modes of labor as productive. By decentering work as the classic human input to production, this perspective may also help us break free from conventional economic analyses that exclude nonhuman animals as laborers. See Henderson, *Value*; Simmel, *The Philosophy of Money*; and Spivak, "Scattered Speculations."
16. There exists a wide-ranging and excellent body of scholarship on the late nineteenth- and early twentieth-century environmental history of the American West to which many of my understandings in this book are indebted. See Cronon, *Nature's Metropolis*; Culver, *The Frontier of Leisure*; Jacoby, *Crimes against Nature*; Langston, *Forest Dreams*; LeCain, *Mass Destruction*; Merrill, *Public Lands*; Sutter, *Driven Wild*; White, *The Organic Machine*; Warren, *The Hunter's Game*; and Worster, *Rivers of Empire*.
17. DeVoto, "The West."
18. In Montana an example of one of these environmental legacies is the toxic Clark Fork River, poisoned by run-off from early twentieth-century mine tailings ponds in the Deer Lodge Valley, not far from the Berkeley Pit, one of the U.S. Environmental

Protection Agency's largest Superfund sites. Local public and private partnerships have worked successfully to clean up these ecological disaster zones for over a decade. But as Tim LeCain points out, while the "mass destruction" of the mining industry may be less prominent in the American West of our own era, it is expanding globally, driven, in part, by American consumers' demands for products such as cell phones that require the exploitation of mineral wealth (*Mass Destruction*, 229–30).

19. My interpretation of these relationships between privatization and enclosure in the modern American West is indebted to an older generation of social historians who suggested that the colonial circumscription of subsistence food practices offered a profound historical shift toward the primitive accumulation of indigenous and work-class means of production. In other words, one way that capital forced people to sell their wages was by criminalizing or forcefully "reforming" traditional modes of sustenance. See Hay, "Poaching"; Thompson, *Whigs and Hunters*; and Wolf, *Europe*. In the case of the American West, Slotkin, *The Fatal Environment*, brilliantly explored corresponding bourgeois fears of industrial workers "going native" with a defiance of wage work in his analysis of late nineteenth-century labor unrest and the Plains Indian Wars.

20. Recent studies have demonstrated that the indirect effects of the presence of wolves on livestock may actually represent a larger economic impact on ranching operations than direct effects of wolf predation. See Steele et al., "Wolf Predation Impacts." It is also possible that a similar effect exists in populations of wild ungulates. For an excellent overview and synthesis of recent studies on the relationship of predation and animal behavior, see Berger, *The Better to Eat You With*.

21. A recent case study on the effect of private property on the history of North American wildlife management is Robert Wilson's excellent book on migratory birds and the wildlife refuges of the U.S. Fish and Wildlife Service, *Seeking Refuge*.

22. See Brower, *Lost Tracks*; Isenberg, *The Destruction*; Smith, *Where Elk Roam*; and Smith, Cole, and Dobkin, *Imperfect Pasture*.

23. See Tuan, *Dominance and Affection*.

1. WOLVES AND WHISKEY

1. This promotional material can be viewed at http://www.russellcountry.com.
2. Marshall, "The Problem of the Wilderness."
3. Stegner, *Wolf Willow*, 7.
4. Flores, *Horizontal Yellow*, 2.

5. For more discussion of Russell's efforts to represent Montana in visual and written mediums, see Cristy, *Charles M. Russell*; Flores, *Visions of the Big Sky*; and Dippie, *Charles M. Russell*.
6. Russell, "Whiskey," 380.
7. See Dempsey, *Firewater*.
8. See Dempsey, *Firewater*; Sharp, *Whoop-Up Country*; and Kennedy, *The Whiskey Trade*.
9. See Overholser, *Fort Benton*.
10. See Sharp, *Whoop-Up Country*, 33–35. For a discussion of steamboat navigation on the Upper Missouri, see Corbin, *The Life and Times*.
11. See Ege, *Tell Baker*, 4.
12. Wissler, "The Social Life," 459.
13. Wissler, "Ceremonial Bundles," 271.
14. See Hungry-Wolf, *The Blackfoot Papers*, 3:810.
15. For more historical detail on the social contexts of nineteenth-century American drinking culture, see Rorabaugh, *The Alcoholic Republic*; and West, *The Saloon*. For an excellent historical discussion of Wissler's ethnographic field research on the Blackfeet Reservation, see Wissler and Kehoe, *Amskapi Pikuni*.
16. Wissler, "Societies and Dance Associations," 436.
17. Davis, "Peyotism," 2.
18. See "The Wolf Man," in Grinnell, *Blackfoot Lodge Tales*, 78–80.
19. Willerslev develops this idea in his brilliant ethnography of Yukaghir hunters in northeastern Siberia. The Yukaghir dress with elk-skin leggings to facilitate the hunt by resembling the sound of elk moving through the undergrowth, what Willerslev sees as an act of mimesis—of mimicking elk. But despite this elk-like behavior, Yukaghir hunters do not believe they actually *become* elk or that they actually become "animals," as any number of other anthropologists operating in colonial paradigms had asserted. Rather, "what defines power in the Yukaghir world," Willerslev argues, "is the ability *not* to confuse analogy with identity . . . the borderland where self and other are both identical and different, alike yet not the same" (*Soul Hunters*, 190). Although different in their specific practices, Blackfoot and Yukaghir understandings of human-animal identities are similar in that their difference as humans is consummated through acts of engagement, through attempts to become animal. Thus, while the Blackfoot are not animal, their concept of natural alliance means that to be human, they are also not *not* animal. See Ingold, *The Perception*; and Taussig, *Mimesis and Alterity*.
20. Bastien, *Blackfoot Ways*, 27.
21. And it presented, as it still does, what Tim Ingold identifies as a "stark choice" of the colonial-genealogical paradigm: "Either we grant indigenous peoples their

historicity, in which case their existence is disconnected from the land, or we allow that their lives are embedded in the land, in which case their historicity is collapsed into an imaginary point of origin.... Land and history, in short, figure as mutually exclusive alternatives" (*The Perception*, 139). The dismantling of Blackfoot natural alliances in the 1870s, both culturally and materially, led to this "stark choice." It rendered Blackfoot historicity in the form of a genealogical kinship that either dissociated Blackfoot people from their environment and removed them from the land or else positioned them, as Tuhiwai Smith might put it, as savages, as timeless, unevolved animals.

22. Bastien, *Blackfoot Ways*, 27.
23. Surgeon's Field Book, Fort Shaw Surgeon's Office Records, SC 1407, MHS.
24. J. H. McKnight & Co., box 4, folder 2, MHS.
25. Senate Executive Document No. 8, 41st Cong., 3rd sess., "Army Posts, Dept. of Dakota," 7, serial 1440, in Sharp, *Whoop-Up Country*, 218.
26. Senate Executive Document No. 57, "Ex-Colonel J. V. D. Reeve," 42nd Cong., 2nd sess. (Washington DC: Government Printing Office, 1872). Like Peter Koch, in the early 1870s Cooper left the Whoop-Up Country and settled in Bozeman, opening a store. Today, Cooper Park sits at the corner of Eighth Avenue and Koch Street.
27. Davison and Tash, "Confederate Backwash."
28. Lepley, *Blackfoot Fur Trade*, 246.
29. Foley, "An Historical Analysis," 23–24.
30. To a degree, this colonial antagonism between the federal government and local settlers was inverted in the Whoop-Up Country compared to contemporary experiences farther west during the 1850s and 1860s. In the case of the Oregon Territory, for instance, historian Gray Whaley has demonstrated that Anglo-American settlers had to pull reluctant federal authorities into participating in colonial wars of Indian removal and extermination, a process Whaley calls "folk imperialism" (*Oregon and the Collapse of Illahee*).
31. This also follows Isenberg's and Flores's arguments that market hunting hastened the destruction of the bison but that Indian nineteenth-century subsistence hunting was also probably unsustainable. In the Whoop-Up Country, killing bison for the market rested almost exclusively in the hands of the Blackfoot; white hunters played a minuscule role in the extermination of the northern herds compared with the more familiar history of the southern and central plains. See Flores, "Bison Ecology"; and Isenberg, *The Destruction*. For more historical background, see Binnema, *Common and Contested Ground*.
32. The chapter "Annihilating Space: Meat" in Cronon, *Nature's Metropolis*, offers a provocative glimpse of this industry's significance and widening geography in the late nineteenth century.

33. *Fort Benton Record*, December 15, 1876, cited in Sharp, *Whoop-Up Country*, 157.
34. Isenberg, *The Destruction*, 100. Isenberg cites Denig, *Five Indian Tribes*, 153; and Lewis, *The Effects of White Contact*, 38–40, 50.
35. Flores, "Bison Ecology," 77.
36. See Whittlesey, "Cows All Over the Place"; and Khan, Mah, and Memish, "Brucellosis in Pregnant Women."
37. Dempsey, *A Blackfoot Winter Count*, 15.
38. See Nugent, "Property Relations."
39. Because of complicated population dynamics and poor historical sources, estimating these populations is difficult and, perhaps, pointless. In his study of the entire North American plains, Isenberg estimates that 1.5 million wolves lived among herds of approximately 30 million bison during the early nineteenth-century peak bison population (*The Destruction*, 26–28). In 1887 William Hornaday estimated that the 1870 bison population north of the Platte River ranged around 1.5 million animals (*The Extermination*, 1887 ed., 504). Given this figure and the proportions stated by Isenberg, the number of plains wolves north of the Platte in 1870 might have been something like 75,000. Defending these figures is challenging, but given other historical evidence, they do seem reasonable. During the early 1870s, the T. C. Power Company alone shipped over 30,000 wolf pelts from the Whoop-Up Country without making a noticeable dent in the overall population. Working backward from Overholser's record of bison robe shipments from 1870 to 1882 and Flores's conservative estimate of a 3 percent net annual bison increase on the southern plains, an 1870 bison population figure for the Whoop-Up Country comes in just shy of a reasonable 450,000. See Bills of Lading for Steamers, 1869–1876, in T. C. Power & Co. Papers, MC 55, box 289, MHS; Flores, "Bison Ecology," 74; and Overholser, *Fort Benton*, 31–32. Applying Isenberg's bison-to-wolf ratio, the baseline 1870 figure for wolves in the Whoop-Up Country would sit at 22,500. See also Fischer, *Wolf Wars*, 14.
40. McGowan, *Animals*, 28, cited in Young, *The Wolf in North American History*, 116.
41. See Fuller, Mech, and Cochrane, "Wolf Population Dynamics."
42. Bills of Lading for Steamers, 1869–1876, T. C. Power & Co. Papers, MC 55, box 289, MHS.
43. From Stanley Young to David Mech, canid biologists have understood this complicated and unexpected relationship between wolf mortality, food availability, and litter size.
44. See Goetzmann, "The Mountain Man." See also Coleman, *Here Lies Hugh Glass*.
45. Summary of Charles Rowe Stories, "Wolves," Overholser Research Files, MAM.
46. Oscar H. Brackett Reminiscences, pp. 0B4–0B5, MHS.
47. Peter Koch in "Letters from the Musselshell," 322, 327, 316.

48. Flores, "Bison Ecology," 78.
49. For a discussion of this massacre, see Goldring, "Whisky, Horses, and Death." See also Graybill, *Policing the Great Plains*, 38–39.
50. Sharp, *Whoop-Up Country*, 52.
51. Sharp, *Whoop-Up Country*, frontispiece.

2. BEASTS OF BOUNTY

1. This is not to say that the politics of public lands and industrial cattle ranching became uncontested by the end of the nineteenth century, just that the industry's logic of predation and production had grown embedded into the broader American economic culture. See Merrill, *Public Lands*.
2. H. A. Riviere, "65 Years Wolver, 1884–1956," 4, Canadian Cattlemen Unpublished Manuscripts, SMF.
3. Mitchell and Greer, "Predators."
4. Montana Stock Growers' Association Executive Committee Meeting, April 17, 1888, Miles City, Montana, Minutes of the Montana Stock Growers' Association, MC 45, 8:178, MHS.
5. Curnow, "The History," 88.
6. Montana Stock Growers' Association Meeting, April 18, 1888, Miles City, Montana, Minutes of the Montana Stock Growers' Association, MC 45, 8:187, MHS.
7. Nelson, "Trapping Wolves," 26.
8. Pallister, "Smoke from the Branding Fire," 40.
9. Long, *Seventy Years*, 338.
10. Riviere, "65 Years Wolver, 1884–1956," 9.
11. Ewing, *The Range*, 198.
12. Ewen Cameron, "Wolves in Montana," 1–2, MC 226, box 6, folder 15, Cameron Papers, MHS.
13. Sixth Legislative Assembly, State of Montana, *Laws, Resolutions and Memorials*, 1899, substitute for House Bill No. 33.
14. Sherm Ewing, interview with Vivian Ellis, 1988, 3, SMF.
15. Coleman, *Vicious*, 10.
16. See Slotkin, *The Fatal Environment*; and Slotkin, *Gunfighter Nation*.
17. For a discussion of this episode, see Evans, *The Bar U*, 119; and Cristy, *Charles M. Russell*, 125.
18. Lears, *Fables of Abundance*, 172.
19. Minutes of the Western Stock Growers' Association meeting, May 11, 1916, Macleod, Alberta, Western Stock Growers' Association Papers, M2452, box 1, folder 2, 29, GLA. Hereafter cited as WSGA plus date and box, folder, and page numbers.

20. These figures come from tabulating thousands of bounty entries in the Montana State Bounty Certificate Records, RS 59, vols. 16–42, 1892–1932 MHS. Special thanks to Rich Aarstad and Jeff Malcomson for hefting each sixty-pound volume in and out of the basement.
21. Thirteenth Legislative Assembly, Territory of Montana, *Laws, Resolutions, and Memorials*, secs. 657–59, art. 3, chap. 26, Fifth Division 1883, 109–10.
22. Western Stock Growers' Association Papers, M2452, box 1, folder 3, GLA.
23. Second Legislative Assembly, State of Montana, *Laws, Resolutions, and Memorials*, 1893, 38–39.
24. Sixth Legislative Assembly, State of Montana, *Laws, Resolutions, and Memorials*, 1899, 100–103.
25. "Length from point of nose to root of tail, four feet; length of tail one foot one inch; height of ear four inches; breadth of ear three inches; from point of nose to end of skull, eleven and one-half inches; from eye to point of nose, four inches," as stated in Hawkins to Rosebud County Sheriff W. M. Moses, September 13, 1913, Montana State Board of Examiners Records, RS 196, box 2, MHS.
26. Bailey, "Key to Animals."
27. Sixth Legislative Assembly, State of Montana, *Laws, Resolutions, and Memorials*, 1899, 100–103.
28. Montana Bounty Certificate Records, RS 59, vol. 23, 1899, MHS.
29. Hawkins to Harold M. Norris, January 13, 1915, Montana State Board of Examiners Records, RS 196, box 2, MHS.
30. Sam Davis to Hawkins, February 24, 1914, and Hawkins to Davis, February 26, 1914, Montana State Board of Examiners Records, RS 196, box 2, MHS.
31. Seventh Legislative Assembly, State of Montana, *Laws, Resolutions, and Memorials*, 1901, 129–30.
32. Montana Bounty Certificate Records, RS 59, vols. 25–39, 1915–1929, MHS.
33. See Broadwater to D. W. Raymond, Secretary of the Montana State Livestock Commission, July 1, 1918, and Walter Brown to Raymond, June 24, 1918, Montana State Board of Examiners Records, RS 196, box 2, MHS.
34. Cameron, "Wolves in Montana," 1–2, MHS.
35. "Foundation for a Wolf Farm," *Helena Daily Herald*, April 4, 1895.
36. Riviere, "65 Years Wolver, 1884–1956," 5.
37. See Breen, *The Canadian Prairie West*; Carter, Evans, and Yeo, *Cowboys, Ranchers, and the Cattle Business*; and Jameson, *Ranches, Cowboys, and Characters*.
38. Elofson, *Cowboys, Gentlemen, and Cattle Thieves*.
39. WSGA, April 1897, box 1, folder 3, GLA.
40. WSGA, August 1897, box 1, folder 3, 30, GLA.
41. WSGA, April 1897, box 1, folder 3, 22–24, GLA.

42. WSGA, September 1900, box 1, folder 3, 98, GLA.
43. WSGA, April 1902, box 1, folder 3, 127, GLA.
44. WSGA, April 1901, box 1, folder 3, 108, GLA.
45. WSGA, May 1908, box 1, folder 3, 188, GLA.
46. WSGA, May 1916, box 1, folder 2, 27–28, GLA.
47. WSGA, May 1916, box 1, folder 2, 27–28, GLA.
48. WSGA, 1915, box 1, folder 3, 236, 240, GLA.
49. WSGA, May 1916, box 1, folder 2, 14–15, GLA.
50. WSGA, May 1916, box 1, folder 2, 14–15, GLA.
51. See Elofson, *Cowboys, Gentlemen, and Cattle Thieves*.

3. MAKING MEAT

1. The exact number of fatalities suffered by the Blackfeet during the 1883–84 starvation winter is unknown. A consensus estimate among historians is six hundred Blackfeet out of a population of no fewer than thirteen hundred. See Farr, *The Reservation Blackfeet*, 8. For an 1883 reservation population estimate, see Inspector Benedict, "Inspection of Blackfeet Agency," July 26, 1883 (3443), in United States Bureau of Indian Affairs, *Reports of the Inspection*, reel 2.
2. Wissler, "The Social Life," 45.
3. The commissioner of Indian Affairs expressed this sentiment directly to Agent Allen in 1886. See J. D. C. Atkins to R. A. Allen, March 15, 1886, in Blackfeet Indian Agency Correspondence—Miscellaneous Letters, 1884–88, RG 75, FARC. Also cited in Samek, *The Blackfoot Confederacy*, 60.
4. Report of Inspector Thomas, "Inspection of Blackfeet agency."
5. See Barsh and Marlor, "Driving Bison"; and Wissler and Kehoe, *Amskapi Pikuni*.
6. Wissler, "Material Culture," 33.
7. Moore, "Cheyenne Work," 129–30.
8. Overholser, *Fort Benton*, 30–32.
9. Nugent, "Property Relations."
10. Ege, *Tell Baker*, 49.
11. Ege, *Tell Baker*, 49.
12. Lubetkin, *Jay Cook's Gamble*, 138–41, 147.
13. Graybill, *Policing the Great Plains*, 51.
14. Ewers, *The Blackfeet*, 239–40.
15. In general, the Blackfeet did not fish, despite living on some of the world's finest trout streams.
16. For a detailed explanation of the peace policy, see Prucha, *American Indian Policy*, 30–71.

17. John Young to the Commissioner of Indian Affairs (CIA), April 10, 1878, in Wessel Papers, box 17, folder 5, MSU.
18. Inspector Howard, "Inspection of the Blackfeet Agency," November 20, 1883 (location no. 4949), United States Bureau of Indian Affairs, *Reports of the Inspection*, reel 2.
19. Young to CIA, July 12, 1881, box 17, folder 8, Wessel Papers, MSU: "The stock cows furnished by the Department will not produce Beef, from its increase, for three or four years, and to prevent outrage and the destruction of the herd something must be done to supply sufficient food."
20. Samek, *The Blackfoot Confederacy*, provides an extensive discussion on the OIA's financial objectives in issuing stock cattle to the Blackfeet in order to scale back rations and cut federal Indian administration expenses.
21. One of the last bison encountered by Blackfeet hunters was killed early in the spring of 1883. See Young to CIA, May 14, 1883, box 18, folder 1, Wessel Papers, MSU.
22. Benedict, "Inspection of Blackfeet Agency," July 26, 1883.
23. H. A. Gillette, M.D., to CIA, letter enclosed in Inspector Howard, "Special Report on the Conditions of the Blackfeet Indians," November 16, 1883 (4862), in United States Bureau of Indian Affairs, *Reports of the Inspection*, reel 2.
24. Howard, "Special Report," November 20, 1883.
25. Agent L. W. Cooke to CIA, "Monthly Report," October 1, 1893, box 19, folder 6, Wessel Papers, MSU.
26. Young to CIA, May 14, 1883, box 18, folder 1; Young to T. C. Power, April 1, 1884, box 18, folder 2; and Young to CIA, "Monthly Report," May 1, 1883, box 18, folder 1, Wessel Papers, MSU. See also Benedict, "Inspection of Blackfeet Agency," July 26, 1883: "By what authority a county sheriff armed with a warrant issued by a territorial district party, enters upon an Indian reservation and makes Indian arrests, and even threatens to arrest the Agent, if he offers any opposition to his mode of procedure, I am unaware. Yet that is the practice in this portion of Montana."
27. Young to CIA, April 10, 1878, box 17, folder 5, Wessel Papers, MSU.
28. Inspector Cisney, "Inspection of Blackfeet Agency," March 25, 1890 (2103), in United States Bureau of Indian Affairs, *Reports of the Inspection*, reel 3.
29. Howard, "Special Report," November 20, 1883; Inspector Barr, "Inspection of Blackfeet Agency," September 22, 1884 (4417); and Inspector Bannister, "Report of Blackfeet Agency," September 29, 1888 (4865), in United States Bureau of Indian Affairs, *Reports of the Inspection*, reel 2.
30. Inspector Gardner, "Inspection Report on the Blackfeet Agency," May 20, 1892 (4372), in United States Bureau of Indian Affairs, *Reports of the Inspection*, reel 3.

31. Cisney, "Inspection of Blackfeet Agency," March 25, 1890.
32. Cooke to CIA, October 1, 1893.
33. Baldwin to CIA, "Monthly Report," September 30, 1888, box 18, folder 6; Baldwin to CIA, "Monthly Report," February 28, 1888, box 18, folder 6, Wessel Papers, MSU.
34. Inspector McCormick, "Report on Blackfeet Agency," September 28, 1893 (7469), in United States Bureau of Indian Affairs, *Reports of the Inspection*, reel 3.
35. Wissler, "Material Culture," 41.
36. McCormick, "Report on Blackfeet Agency," September 28, 1893.
37. Cisney, "Inspection of Blackfeet Agency," March 25, 1890.
38. Farr, *The Reservation Blackfeet*, 14–15, 52.
39. Howard, "Special Report," November 20, 1883.
40. Cooke, "Monthly Report," December 1, 1893.
41. Inspector Howard, "Special Report on Conditions of Blackfeet Indians," November 16, 1883 (4862), in United States Bureau of Indian Affairs, *Reports of the Inspection*, reel 2.

4. THE PLACE THAT FEEDS YOU

1. Stegner, *Wolf Willow*, 282–83.
2. I do not speak Blackfoot. For my reconstructions of Blackfoot words and concepts, including *áuasini*, I have relied mainly on two sources: Bastien, *Blackfoot Ways*, 197–98; and Uhlenbeck, *An English-Blackfoot Vocabulary*, 54.
3. "Minutes of Meeting: Resolutions Committee," March 30, 1925, Record Group 75, CCF-1, file 27506-1923-BF-100, National Archives and Records Administration (NARA), published in Rosier, *Rebirth*, 36.
4. See Hoxie, *A Final Promise*; Genetin-Pilawa, *Crooked Paths to Allotment*; and Prucha, *The Great Father*.
5. See Lears, *Fables of Abundance*; and Livingston, *Pragmatism, Feminism, and Democracy*.
6. See Bantjes, *Improved Earth*; Fiege, "The Weedy West"; and Fitzgerald, *Every Farm a Factory*.
7. See Ewers, *The Blackfeet*; Rosier, *Rebirth*; and Wessel, "Agriculture on the Reservations."
8. Rosier, *Rebirth*, 34.
9. Agent James Monteath to Commissioner of Indian Affairs, October 16, 1902, file 78824, Records of the Bureau of Indian Affairs, Record Group 75, NARA, cited in Wessel, "Agriculture on the Reservations," 18–23.
10. George Steell to Commissioner of Indian Affairs, February 27, 1893, Record Group 75, Federal Records Center, Denver, Colorado, cited in Wessel, "Agriculture on the Reservations."

11. Steell to Herman Knoell, Flowerree and Lowery Cattle Co., August 6, 1892, box 19, folder 4, Wessel Papers, MSU.
12. See Inspector Gardner to CIA, May 14, 1892 (3527), in United States Bureau of Indian Affairs, *Reports of the Inspection*, reel 3.
13. Clarence Churchill to Auditor, U.S. Treasury Department, January 4, 1907, in box 2, folder 4, Wessel Papers, MSU.
14. See J. Z. Dare to CIA, April 26, 1907, box 2, folder 2; James Sanders to CIA, September 3, 1908, box 2, folder 3; Churchill to CIA, April 23, 1909, box 2, folder 4; McFatridge to CIA, November 28, 1911, box 3, folder 2; and Abbot to Secretary of Interior, January 22, 1912, box 3, folder 3, all in Wessel Papers, MSU.
15. Churchill to CIA, September 15, 1909, box 2, folder 4, Wessel Papers, MSU.
16. J. Z. Dare to Auditor, U.S. Treasury Department, January 4, 1907, box 2, folder 2, Wessel Papers, MSU.
17. For a discussion of Blackfeet allotment selections, see Farr, *The Reservation Blackfeet*.
18. Churchill to CIA, September 15, 1909, box 2, folder 4, Wessel Papers, MSU.
19. Dare to CIA, May 1, 1907, box 2, folder 2, Wessel Papers, MSU.
20. "Annual Report of the Blackfeet Agency," November 12, 1909, box 2, folder 4, Wessel Papers, MSU.
21. Samek, *The Blackfoot Confederacy*, 51.
22. "Annual Report of the Blackfeet Agency," November 12, 1909.
23. In Evelyn M. Rhoden, "Cattle Rustling," box 53, folder 8; and "Blackfoot Cattle Brands," box 73, folder 6, both in WPA Records, Collection 2336, MSU.
24. "Annual Report of the Blackfeet Agency," November 12, 1909.
25. See Milner and O'Connor, *As Big as the West*; and Brown, "Violence," 401–2.
26. See McFatridge to CIA, April 16, 1912, in box 3, folder 4, Wessel Papers, MSU.
27. Churchill to CIA, September 15, 1909.
28. See Ewers, "Trading Land for a Living," in Ewers, *The Blackfeet*.
29. Churchill to CIA, July 12, 1909, in box 2, folder 4, Wessel Papers, MSU.
30. Rosier, *Rebirth*, is one such work that analyzes Blackfeet tribal history on the basis of "full blood" and "mixed blood" as cultural and political categories.
31. Oscar Lipps, Supervisor, "Report on the Education, Industrial, Economic, and Home Conditions, Blackfeet Reservation, Montana," October 20, 1913, box 4, folder 1, Wessel Papers, MSU.
32. Lipps, "Report on the Education."
33. Dare to CIA, October 4, 1906, box 2, folder 1, Wessel Papers, MSU.
34. See Churchill to CIA, July 12, 1909. Blackfeet opponents of allotment did not necessarily view privatization itself as a means of dispossession, only the sale of surplus lands. See Churchill, "Transcript of Tribal Council Meeting at Blackfeet

Agency," April 24, 1909, box 2, folder 4, Wessel Papers, MSU: "The Indians are very anxious to have the allotting completed so that they can make permanent improvements and establish suitable homes. They have been looking forward to the time when allotments would be made them for about ten years. While there have been several instances of personal differences arising from disputes over selections for allotment, there have been no general complaints nor individual opposition to allotment." In the future, I plan to further analyze Blackfeet debates over allotment in comparison the earlier allotment of the "closed reservations" of the Blackfoot, Blood, and Piegan Reserves in Alberta. See Samek, *The Blackfoot Confederacy*.

35. See Rosier, *Rebirth*.
36. See McFatridge to CIA, April 17, 1914, box 4, folder 2; McFatridge to CIA, January 31, 1914, box 4, folder 2; and McFatridge to CIA, April 16, 1914, box 4, folder 2, all in Wessel Papers, MSU.
37. See McFatridge to CIA, January 19, 1914, box 4, folder 2; McFatridge to CIA, January 29, 1914, box 4, folder 2; McFatridge to CIA, January 31, 1914, box 4, folder 2; and Robert Hamilton to A. R. Serven, January 31, 1914, box 4, folder 2, all in Wessel Papers, MSU.
38. Rosier, *Rebirth*, 43.
39. See Wheeler, *Yankee from the West*, 177.
40. Wessel, "Agriculture on the Reservations," 17.
41. Greenwald, *Reconfiguring*, 145.
42. Chang, *The Color of the Land*.
43. See Rosier, *Rebirth*.
44. For a good description of these Blackfeet concepts, see Denman, "Cultural Change." The phrase "self-torture" may seem jarring to some readers, but this was, in fact, the purpose of specific Sun Dance rituals enacted by the Blackfeet and other Plains peoples who understood these painful experiences as methods of personal transformation.
45. See Fred Campbell, "Five-Year Industrial Program from April 1, 1921 to April 1, 1926," box 9, folder 1, Wessel Papers, MSU.
46. See Wessel, "Agriculture on the Reservations," for background on the Monteath relocation.
47. Fred Campbell, "The Five-Year Program on the Blackfeet Indian Reservation," *Indian Leader* 26, no. 25 (1923): 3.
48. For an analysis of agricultural cycles of debt, equipment investment, and overproduction, see Fitzgerald, *Every Farm a Factory*.
49. "Extract of the Report of General H. L. Scott, Dated October 10, 1925," box 8, folder 6, Wessel Papers, MSU.

50. Oscar Lipps, "Subsistence Farming on the Blackfeet Reservation; Memorandum for Mr. Cooley," box 8, folder 5, Wessel Papers, MSU.
51. Campbell, "The Five-Year Program," 3.
52. Schultz, *The Starving Blackfeet Indians*.
53. "Extract of the Report of General H. L. Scott."

5. UNNATURAL HUNGER

1. "Kill Buffalo," *Minneapolis Tribune*, December 31, 1900, 7.
2. I do not have adequate space in this chapter to fully develop the significance of these narratives of predation in structuring current debates over bison conservation. However, one noteworthy example is the brucellosis controversy on the Yellowstone winter range. While ecologists agree that a lack of sufficient winter habitat within the park is the limiting factor in the health of Yellowstone's migratory bison herds, public outcry against Montana's recreational bison hunters is far more visible than the ongoing efforts to expand winter bison habitat on private lands adjacent to the park. In part, this chapter tries to explain why it is easier to sell to the liberal public a discourse of antihunting than it is to push for the repatriation of former wildlife habitat.
3. For a classic statement of the federal legal framework on conservation that emerged in the early twentieth century, see Hays, *Conservation*; and Nash, *Wilderness*.
4. See A. Brazier Howell to Paul Redington, undated, CONS 90, box 3, folder 33, Murie Papers, DPL.
5. Stanley Young to Arthur Carhart, November 24, 1930, CONS 83, box 8, folder 40, Young Papers, DPL.
6. William Hornaday, "Report of the President on the Founding of the Montana National Bison Herd," in *The Second Annual Report of the American Bison Society* (New York: American Bison Society, 1909), 1–18, quote from 1.
7. "The Vanishing Buffalo," *Minneapolis Morning Tribune*, February 20, 1910, 54, reprinted from the *Washington Post*.
8. "National Bison Preserve Turns History Backward," *Minneapolis Morning Tribune*, February 5, 1911, 14.
9. "The Vanishing Buffalo."
10. Trosper, *The Economic Impact*, 35.
11. For more on Samuel Walking Coyote and the history of the Pablo-Allard bison herd, see Whealdon, *"I Will Be Meat"*; and Zontek, *Buffalo Nation*.
12. Hornaday, "Report of the President," 5.
13. Hornaday, "Report of the President," 9.
14. For a discussion of the history of experiments with cattle-bison hybrids, see Isenberg, *The Destruction*.

15. William Hornaday to Morton J. Elrod, June 30, 1908, box 1, folder 2, American Bison Society Papers, DPL.
16. For details on the James J. Hill bison saga, see William T. Hornaday to Dr. Palmer, January 7, 1909; William T. Hornaday to John Toomey, January 12, 1909; William T. Hornaday to James J. Hill, January 12, 1909; William T. Hornaday to James J. Hill, October 18, 1909; William T. Hornaday to James J. Hill, August 30, 1910; William T. Hornaday to James J. Hill, October 14, 1910, all in box 1, folder 2, CONS 4, American Bison Society Papers, DPL.
17. Hornaday, "Report of the President," 13.
18. Hornaday, *The Extermination*, 1887 ed.
19. For a discussion of Hornaday's trip to the Judith Basin, see Hanna Rose Shell's foreword in the 2002 edition of Hornaday, *The Extermination*. Also see Barrow, *Nature's Ghosts*.
20. William Hornaday to Carl Hagenbeck, June 11, 1902, cited in Rothfels, *Savages and Beasts*, 67.
21. Hornaday to Frederic H. Kennard, January 23, 1909, box 1, folder 2, American Bison Society Papers, DPL.
22. Hornaday to Jimmy Simpson, April 4, 1912, M78, folder 3, James Simpson Family Fonds, Whyte Museum, Banff, Alberta.
23. See Jacoby, *Crimes against Nature*; and Warren, *The Hunter's Game*.
24. Hornaday, *Our Vanishing Wildlife*, 101.
25. Hornaday, "The Destruction," 77.
26. See Isenberg, "The Returns of the Bison."
27. This is how Brinkley, *Wilderness Warrior*, 284, puts it.
28. William T. Hornaday, *The Annual Report of the New York Zoological Society* (New York, 1916), 20:66.
29. Hornaday, "Report of the President," 13.
30. Hornaday to W. P. Wharton, July 20, 1908, box 1, folder 2, American Bison Society Papers, DPL.
31. Hornaday to Frederic Kennard, April 5, 1909, box 1, folder 2, American Bison Society Papers, DPL.
32. Hornaday to W. G. Nye, Commissioner of Public Affairs of the Commercial Club, Minneapolis, October 6, 1908, box 1, folder 2, American Bison Society Papers, DPL.
33. "Wants City to Buy a Buffalo," *Minneapolis Tribune*, September 26, 1908, 6.
34. Hornaday to Fred B. Strunz, November 23, 1908, box 1, folder 2, American Bison Society Papers, DPL.
35. Hornaday to Howard Elliot, February 8, 1909, box 1, folder 2, American Bison Society Papers, DPL.

36. "End of Famous Frontiersman," *Helena Independent*, July 18, 1901.
37. "Buffalo Hunt by Auto Ends in Beast's Death," *Minneapolis Morning Tribune*, December 17, 1910.
38. See the following newspaper articles: "Auto Eclipses Buffalo in Exciting Anoka County Chase," *Minneapolis Morning Tribune*, December 18, 1910; "Farmers Take Down Guns to Fight Warring Bison," *Minneapolis Morning Tribune*, December 16, 1910; "Buffalo Scare at Anoka Rouses Town and Country," *Minneapolis Morning Tribune*, December 14, 1910; and "Speedy Auto Takes Bronco's Place in Buffalo Chase," *Minneapolis Morning Tribune*, December 18, 1910.
39. Hornaday to James J. Hill, August 30, 1910, box 1, folder 2, American Bison Society Papers, DPL.
40. Hornaday to James J. Hill, October 14, 1910, box 1, folder 2, American Bison Society Papers, DPL.
41. For a detailed analysis of the frontier idea and the formation of racial identities in the early twentieth-century American West, see Klein, *Frontiers*.
42. "Old Times and New," *Stockgrowers Journal*, Miles City, Montana, July 10, 1907, box 8, folder 28, L. A. Huffman Studio Collection, BBHC.
43. L. A. Huffman, "Mex John Making Pies," undated photograph, Montana State Historical Society, Helena, Montana. Also online at http://digitalarchive.oclc.org/request?id=oclcnum:70586358.
44. Edith Franz to L. A. Huffman (LAH), undated, box 5, folder 1; Badger Clark to LAH, January 4, 1921, box 2, folder 40; and John E. Steele to LAH, undated, box 1, folder 24, all in L. A. Huffman Studio Collection, BBHC.
45. These requests are too numerous to count. See L. A. Huffman Studio Collection, BBHC.
46. "Proceedings around the Welcome Camp-Fire Given in Honor of Colonel Theodore Roosevelt at the Waldorf-Astoria," Camp-Fire Club of America, June 22, 1910, in box 4, folder 24, L. A. Huffman Studio Collection, BBHC.
47. See the many correspondences between Edmund Seymour and LAH regarding oil in box 1, folders 21 and 22, L. A. Huffman Studio Collection, BBHC.
48. See George Washington Moore to J. E. Campbell, December 11, 1923, box 3, folder 69, L. A. Huffman Studio Collection, BBHC. More evidence on the impending implosion of the company can be found in Black Panther Oil and Refining Corporation, "Financial Statement and Report," July 30, 1921, box 6, folder 11, L. A. Huffman Studio Collection, BBHC.
49. Information on the Montana Mutual Oil Syndicate is available in the many manuscripts held in box 6, folder 46, L. A. Huffman Studio Collection, BBHC. In particular, see Eugene Kegley to George Roberts, November 25, 1921, box 6, folder 46, L. A. Huffman Studio Collection, BBHC. Hornaday was the exception.

In 1921 Hornaday responded to Huffman's investment suggestion, explaining that "having recently contributed a modest $250 to a dry hole in Oklahoma, which a shrewd and very dear friend became convinced was a sure thing, I am sore on all well drilling. My friends are all dry hole investors." Hornaday might have been right with his arithmetic when he opined to Huffman that "incidentally, it is my belief that two dollars go down into the earth in dry holes for every one that comes out in the shape of oil."

50. See Turner, "The Significance of the Frontier."
51. Edmund Seymour to L. A. Huffman, undated, box 4, folder 10, American Bison Society Papers, DPL.
52. Leonard and Holmes, *Who's Who*, 1170.
53. Edmund Seymour to A. A. Sugg, August 5, no year date, MSS 1, box 14, folder 2, William Hornaday Papers, Wildlife Conservation Society Archives, Bronx Zoo, New York (WCS).
54. Seymour to M. S. Carpenter, July 16, 1926, MSS 1, box 14, folder 2, WCS.
55. Seymour to Hornaday, December 20, 1919 box 3, folder 24, American Bison Society Papers, DPL. "It comes to me as a rumor that your friend Grant put his foot in it out in Oregon . . . offering a reward of $200 himself for the conviction of anybody killing antelope."
56. I am referring to a natural history paradigm that emphasized the habits and characteristics of organisms as predetermined taxonomic certainties rather than as mutable social and ecological behaviors.
57. On these misuses of Darwin, see Gould, *The Mismeasure of Man*. See also Grant, *The Passing*; and Ripley, *The Races of Europe*.
58. Here is Hornaday's explanation and justification for his methodology: "Any naturalist who is worth his salt can determine the thoughts and feelings of mammals, birds, and reptiles by the sign language, by facial and bodily expression, and by telepathy; and afterward he can easily translate the whole interview product into outdoor English for the benefit of the reading classes" (*Wild Animal Interviews*, ix).
59. Antelope; Tiger; Wolf. See Hornaday, *Wild Animal Interviews*, 163, 235, 295.
60. Quoted in Bigelow, "Frederic Remington," and cited in White, *The Eastern Establishment*, 109.
61. Few historians have explored this link between the new nativism and Indian dispossession better than Richard Slotkin, who plotted the Indian Wars in relation to eastern labor disputes following the Civil War. See Slotkin, *The Fatal Environment*.
62. "Speeding to the Rockies," *Washington Post*, April 14, 1905, 1, cited in Brinkley, *Wilderness Warrior*, 612.

63. George Bird Grinnell to LAH, October 28, 1908, box 10, folder 64, L. A. Huffman Studio Collection, BBHC.
64. While there is not yet a comprehensive history of Basque immigration to Montana during the Progressive Era, Etulain and Echeverria, *Portraits of Basques*, offers a useful selection of historical articles on Basque migration more generally.
65. Hornaday, "Diversions."
66. Hornaday, "Diversions," 1.
67. Hornaday, "Diversions," 17.
68. Henry Fairfield Osborn, preface in Grant, *The Passing*, ix.
69. Grant, *The Passing*, 234–35.
70. *The Annual Report of the American Bison Society* (1922), 41, 63, in box 1, folder 23, L. A. Huffman Studio Collection, BBHC.
71. Van Nuys, *Americanizing the West*, 192.
72. Brower, *Lost Tracks*, 94.
73. See Carhart and Young, *The Last Stand*. Also see Coleman, *Vicious*, for an account of this book's development, as well as Young's early partnership with Carhart.
74. In 1939 the Bureau of Biological Survey and Fisheries turned into the U.S. Fish and Wildlife Service. Carhart was not the first to question the Bureau of Biological Survey's predator control mission. Several outspoken members of the American Society of Mammalogists, including Charles Adams, Joseph Grinnell, and E. Raymond Hall, along with other affiliates of the University of California's Museum of Vertebrate Zoology, had been pressing the bureau about its predator control agenda since the mid-1920s. See Dunlap, *Saving America's Wildlife*, and Robinson, *Predatory Bureaucracy*, for excellent analyses of this debate.
75. Young to Carhart, November 24, 1930.
76. Young to Carhart, November 24, 1930.
77. A. Brazier Howell to Paul Redington, undated, CONS 90, box 3, folder 33, Murie Papers, DPL.
78. Dunlap, *Saving America's Wildlife*, and Worster, *Nature's Economy*, both give accounts of the scientific battles between the ASM and PARC before and after the passage of the Animal Damage Control Act.
79. See Dunlap, *Saving America's Wildlife*; and Robinson, *Predatory Bureaucracy*.
80. See the Animal Damage Control Act of March 2, 1931.
81. Young to Carhart, November 24, 1930.
82. Brower, *Lost Tracks*, 60.
83. Hornaday to Hooper, June 16, 1913.
84. Menu inserted in *The Sixth Annual Report of the American Bison Society* (New York, 1913), American Bison Society Papers, DPL.

85. Northern Pacific advertisement, box 6, folder 38, American Bison Society Papers, DPL. Also cited in Isenberg, *The Destruction*, 184.

CONCLUSION

1. See Berger, *The Better to Eat You With*, 3, 38.
2. See "Wolves Devastate Ranchers Sheep," *Billings Gazette*, August 27, 2009.
3. United States Department of Fish and Wildlife, *The Reintroduction of Gray Wolves*, chap. 5, "Consultation and Coordination," 51.
4. Wolch and Emel, *Animal Geographies*, 112.
5. Cameron, "Wolves in Montana," 9, 11.
6. Haraway provocatively asks: "How can remembering the conquest of the western states by Anglo settlers and their plants and animals become part of the solution and not another occasion for the pleasurable and individualizing frisson of guilt?" Her self-satisfying solution is to "reach out and pet one's dog" (Haraway, *When Species Meet*, 41).
7. See Mowat, *Never Cry Wolf*; and Carol Ballard, *Never Cry Wolf* (Los Angeles: Walt Disney Productions, 1983).
8. Werner Herzog, *Grizzly Man* (Santa Monica: Lion's Gate Entertainment, 2005).
9. Graham and Beard, *Eyelids of Morning*, 12.
10. Petersen, *Heartsblood*, 71.

Bibliography

ARCHIVES AND MANUSCRIPT MATERIALS

Buffalo Bill Historical Center, Cody, Wyoming (BBHC)
    L. A. Huffman Studio Collection, McCracken Research Library
Denver Public Library, Denver, Colorado (DPL)
    American Bison Society Papers, Conservation Collection
    Olaus J. Murie Papers, Conservation Collection
    Stanley Paul Young Papers, Conservation Collection
Glenbow Library and Archives, Calgary, Alberta (GLA)
    David C. Duvall Fonds, Archives
    Jane and Lucien Hanks Fonds, Archives
    New Waldron Ranch Fonds, Archives
    Western Stock Growers' Association Papers
Montana Agricultural Museum, Fort Benton, Montana (MAM)
    Overholser Research Files, Overholser Library
Montana Historical Society Research Center, Helena, Montana (MHS)
    Evelyn and Ewen Cameron Papers, Archives
    Fort Shaw Surgeon's Office Records, Archives
    J. H. McKnight and Company Records, Archives
    Minutes of the Montana Stock Growers' Association, Archives
    Montana State Board of Examiners Records, Archives
    Montana State Bounty Certificate Records, Archives
    Oscar H. Brackett Reminiscences, Archives
    T. C. Power and Company Records, Archives
Montana State University, Bozeman, Montana (MSU)

Montana Writer's Project Records, Burlingame Special Collections
   Thomas R. Wessel Indian Claims Commission Research Papers, 1855–1979, Collection 2059, Burlingame Special Collections
   WPA Records, Collection 2336
Newberry Library, Chicago, Illinois
   D'Arcy McNickle Papers, Baskes Special Collections
   Rocky Mountain Region Federal Archives and Records Center, Denver, Colorado (FARC)
   Blackfeet Indian Agency Correspondence—Miscellaneous Letters, 1884–88
Stockmen's Memorial Foundation, Cochrane, Alberta (SMF)
   Canadian Cattlemen Unpublished Manuscripts, Bert Sheppard Library and Archives
   Sherm Ewing Oral History Collection, Bert Sheppard Library and Archives
University of Montana, Missoula, Montana
   Sherburne Family Papers, Mansfield Library Special Collections
Whyte Museum, Banff, Alberta
   James Simpson Family Fonds, Archives
Wildlife Conservation Society Library, Bronx, New York
   William Hornaday Papers, Archives

PUBLISHED WORKS

American Bison Society. *Annual Reports of the American Bison Society*. New York: American Bison Society, 1905–1923.
Bailey, Vernon. "United States Department of Agriculture, Bureau of Biological Survey—Circular No. 60: Key to Animals on which Wolf and Coyote Bounties Are Often Paid." Washington DC: Government Printing Office, 1909.
Bantjes, Rod. *Improved Earth: Prairie Space as Modern Artefact, 1869–1944*. Toronto: University of Toronto Press, 2005.
Barrow, Mark. *Nature's Ghosts: Confronting Extinction from the Age of Jefferson to the Age of Ecology*. Chicago: University of Chicago Press, 2009.
Barsh, Russell Lawrence, and Chantelle Marlor. "Driving Bison and Blackfoot Science." *Human Ecology* 31, no. 4 (December 2003): 571–93.
Bastien, Betty. *Blackfoot Ways of Knowing: The Worldview of the Siksikaitsitapi*. Calgary: University of Calgary Press, 2004.
Bederman, Gail. *Manliness and Civilization: A Cultural History of Gender and Race in the United States, 1880–1917*. Chicago: University of Chicago Press, 1995.
Bell, W. B. "Hunting Down Stock Killers." Separate No. 845, in *Yearbook of the United States Department of Agriculture*, U.S. Bureau of Biological Survey. Washington DC: Government Printing Office, 1920.

Berger, Joel. *The Better to Eat You With: Fear in the Animal World*. Chicago: University of Chicago Press, 2008.

Bigelow, Poultney. "Frederic Remington; with Extracts from Unpublished Letters." *New York State Historical Association Quarterly Journal* 10 (1929): 46–48.

Binnema, Theodore. *Common and Contested Ground: A Human and Environmental History of the Northwestern Plains*. Norman: University of Oklahoma Press, 2001.

Brechin, Gray. "Conserving the Race: Natural Aristocracies, Eugenics, and the U.S. Conservation Movement." *Antipode* 28, no. 3 (1996): 229–45.

Breen, David. *The Canadian Prairie West and the Ranching Frontier: 1874–1924*. Toronto: University of Toronto Press, 1983.

Brinkley, Douglas. *Wilderness Warrior: Theodore Roosevelt and the Crusade for America*. New York: HarperCollins, 2009.

Brower, Jennifer. *Lost Tracks: Buffalo National Park, 1909–1939*. Edmonton: Athabasca University Press, 2008.

Brown, Richard Maxwell. *No Duty to Retreat: Violence and Values in American History and Society*. Norman: University of Oklahoma Press, 1994.

Buck, Fred E. "Survey Party on the Blackfeet Indian Reservation, 1907." *Montana: The Magazine of Western History* 34, no. 1 (1984): 57–61.

Carhart, Arthur, and Stanley Young. *The Last Stand of the Pack*. New York: J. H. Sears and Company, 1929.

Carter, Sarah, Simon Evans, and Bill Yeo. *Cowboys, Ranchers, and the Cattle Business: Cross-Border Perspectives on Ranching History*. Boulder: University of Colorado Press, 2000.

Cartmill, Matt. *A View to a Death in the Morning: Hunting and Nature through History*. Cambridge MA: Harvard University Press, 1993.

Chang, David. *The Color of the Land: Race, Nation, and the Politics of Landownership in Oklahoma, 1832–1929*. Chapel Hill: University of North Carolina Press, 2010.

Child, Brenda. *Boarding School Seasons: American Indian Families, 1900–1940*. Lincoln: University of Nebraska Press, 2000.

Coleman, Jon T. *Here Lies Hugh Glass: A Bear, a Mountain, and the Rise of the American Nation*. New York: Hill & Wang, 2012.

———. *Vicious: Wolves and Men in America*. New Haven, CT: Yale University Press, 2004.

Collingham, E. M. *Imperial Bodies: The Physical Experience of the Raj, c. 1800–1947*. Cambridge: Polity Press, 2001.

Corbin, Annalies. *The Life and Times of the Steamboat Red Cloud, or, How Merchants, Mounties, and the Missouri Transformed the West*. College Station: Texas A&M University Press, 2006.

Cristy, Raphael. *Charles M. Russell: The Storyteller's Art*. Albuquerque: University of New Mexico Press, 2004.
Cronon, William. *Nature's Metropolis: Chicago and the Great West*. 1st ed. New York: W. W. Norton, 1991.
———. "The Trouble with Wilderness: Or, Getting Back to the Wrong Nature." *Environmental History* 1, no. 1 (1996): 7–28.
Culver, Lawrence. *The Frontier of Leisure: Southern California and the Shaping of Modern America*. Oxford: Oxford University Press, 2010.
Curnow, Edward M. "The History of the Eradication of the Wolf in Montana." MA thesis, University of Montana, 1969.
Davis, Leslie B. "Peyotism and the Blackfeet Indians of Montana: An Historical Assessment." *Studies in Plains Anthropology and History* 1 (1961). Browning: Museum of the Plains Indian and U.S. Office of Indian Affairs.
Davis, Mike. *Late Victorian Holocausts: El Niño Famines and the Making of the Third World*. London: Verso, 2001.
Davison, Stanley R., and Dale Tash. "Confederate Backwash in Montana Territory." In *The Montana Past: An Anthology*, edited by Michael P. Malone and Richard B. Roeder, 111–20. Missoula: University of Montana Press, 1969.
Dehler, Gregory John. "An American Crusader: William Temple Hornaday and Wildlife Protection in America, 1840–1940." PhD diss., Lehigh University, 2002.
Deloria, Philip Joseph. *Indians in Unexpected Places*. Lawrence: University Press of Kansas, 2004.
Dempsey, Hugh A., ed. *A Blackfoot Winter Count*. Calgary: Glenbow Foundation, 1965.
———. *Firewater: The Impact of the Whisky Trade on the Blackfoot Nation*. Calgary: Fifth House Publishers, 2002.
Denig, Edward. *Five Indian Tribes of the Upper Missouri: Sioux, Arikaras, Assiniboines, Crees, Crows*. Edited by John C. Ewers. Norman: University of Oklahoma Press.
Denman, Clayton C. "Cultural Change among the Blackfeet Indians of Montana." PhD diss., University of California, Berkeley, 1968.
Derrida, Jacques. "The Animal That Therefore I Am (More to Follow)." Translated by David Wills. *Critical Inquiry* 28 (Winter 2002): 369–418.
DeVoto, Bernard. "The West: A Plundered Province." *Harper's Monthly*, August 1934, 335–64.
Dippie, Brian, ed. *Charles M. Russell, Word Painter: Letters, 1877–1926*. Fort Worth: Amon Carter Museum, 1993.
Duerr, Hans Peter. *Dreamtime: Concerning the Boundary between Wilderness and Civilization*. Translated by Felicitas Goodman. 1978; Oxford: Basil Blackwell, 1985.
Dunlap, Thomas R. *Saving America's Wildlife: Ecology and the American Mind, 1850–1990*. Princeton NJ: Princeton University Press, 1988.

Ege, Robert. *Tell Baker to Strike Them Hard: Incident on the Marias, 23 Jan. 1870*. Fort Collins: Old Army Press, 1970.

Elofson, Warren. *Cowboys, Gentlemen, and Cattle Thieves*. Montreal: McGill-Queen's University Press, 2000.

———. "An Exceedingly Dicey Business: Frontier Horse Ranching on the Northern Great Plains." *Agricultural History* 79, no. 4 (2005): 462–77.

———. "Law and Disorder on the Ranching Frontiers of Montana and Alberta/Assiniboia, 1870–1914." *Journal of the West* 42, no. 1 (2003): 40–51.

Elson, Diane. "A Value Theory of Labor." In *Value: The Representation of Labour in Capitalism*, edited by Diane Elson, 115–80. London: CSE Books, 1979.

Etulain, Richard W., and Jeronima Echeverria, eds. *Portraits of Basques in the New World*. Reno: University of Nevada Press, 1999.

Evans, Simon M. *The Bar U and Canadian Ranching History*. Calgary: University of Calgary Press, 2004.

Evans, Sterling, ed. *The Borderlands of the American and Canadian Wests: Essays on Regional History of the Forty-Ninth Parallel*. Lincoln: University of Nebraska Press, 2006.

Ewing, Sherm. *The Range*. Missoula: Mountain Press Publishing, 1991.

Farr, William. *The Reservation Blackfeet: A Photographic History of Cultural Survival*. Seattle: University of Washington Press, 1984.

Fiege, Mark. "The Weedy West: Mobile Nature, Boundaries, and Common Space in the Montana Landscape." *Western Historical Quarterly* 35, no. 1 (Spring 2005): 22–47.

Fischer, Hank. *Wolf Wars: The Remarkable Inside Story of the Restoration of Wolves to Yellowstone*. Helena MT: Falcon Press, 1995.

Fitzgerald, Deborah. *Every Farm a Factory: The Industrial Ideal in American Agriculture*. New Haven CT: Yale University Press, 2003.

Flores, Dan L. *Horizontal Yellow: Nature and History in the Near Southwest*. Albuquerque: University of New Mexico Press, 1999.

———. *The Natural West: Environmental History in the Great Plains and Rocky Mountains*. Norman: University of Oklahoma Press, 2001.

———. *Visions of the Big Sky: Painting and Photographing the Northern Rocky Mountain West*. Norman: University of Oklahoma Press, 2010.

Foley, Michael. "An Historical Analysis of the Administration of the Blackfeet Indian Reservation by the United States, 1855–1950s." Indian Claims Commission, Docket No. 279-D.

Forster, Elborg, and Patricia Ranum, eds. *Food and Drink in History: Selections from the "Annales. Economies, sociétés, civilisations."* Baltimore MD: Johns Hopkins University Press, 1979.

Fradkin, Philip L. *Wallace Stegner and the American West*. 1st ed. New York: Alfred A. Knopf, 2008.

Fuller, Todd K., L. David Mech, and Jean Fitts Cochrane. "Wolf Population Dynamics." In *Wolves: Behavior, Ecology, and Conservation*, edited by L. David Mech and Luigi Boitani, 182–85. Chicago: University of Chicago Press, 2003.

Genetin-Pilawa, C. Joseph. *Crooked Paths to Allotment: The Fight over Federal Indian Policy after the Civil War*. Chapel Hill: University of North Carolina Press, 2012.

Goetzmann, William. "The Mountain Man as Jacksonian Man." *American Quarterly* 15, no. 3 (Autumn 1963): 402–15.

Goldring, Philip. "Whisky, Horses, and Death." *Canadian West* 7, no. 4 (1991): 140–47.

Gould, Stephen Jay. *The Mismeasure of Man*. New York: W. W. Norton, 1981.

Graham, Alistair, and Peter Beard. *Eyelids of Morning: The Mingled Destinies of Crocodiles and Men*. 1973; San Francisco: Chronicle Books, 1990.

Grant, Madison. *The Passing of the Great Race: Or, the Racial Basis of European History*. 3rd ed. New York: Scribner's, 1919.

Graybill, Andrew R. *Policing the Great Plains: Rangers, Mounties, and the North American Frontier, 1875–1910*. Lincoln: University of Nebraska Press, 2007.

———. *The Red and the White: A Family Saga of the American West*. New York: W. W. Norton, 2013.

Greenwald, Emily. *Reconfiguring the Reservation: The Nez Perces, Jicarilla Apaches, and the Dawes Act*. Lincoln: University of Nebraska Press, 2002.

Grinnell, George Bird. *Blackfoot Lodge Tales*. New York: Scribner's, 1920.

Gutfeld, Arnon. "Western Justice and the Rule of Law: Bourquin on Loyalty, the 'Red Scare,' and Indians." *Pacific Historical Review* 65, no. 1 (1996): 85–106.

Haraway, Donna Jeanne. *When Species Meet*. Minneapolis: University of Minnesota Press, 2008.

Harkin, Michael Eugene, and David Rich Lewis, eds. *Native Americans and the Environment: Perspectives on the Ecological Indian*. Lincoln: University of Nebraska Press, 2007.

Hassrick, Peter H. "'They Are a Fine Outfit Those Blackfeet': Frederic Remington in Western Canada." *Alberta History* 52, no. 2 (2004): 27–48.

Hay, Douglas. "Poaching and the Game Laws on Cannock Chase." In *Albion's Fatal Tree: Crime and Society in Eighteenth-Century England*, 189–254. New York: Pantheon, 1975.

Hays, Samuel P. *Conservation and the Gospel of Efficiency: The Progressive Conservation Movement, 1890–1920*. Cambridge MA: Harvard University Press, 1959.

Henderson, George L. *California and the Fictions of Capital*. New York: Oxford University Press, 1999.

———. *Value in Marx: The Persistence of Value in a More-Than-Capitalist World*. Minneapolis: University of Minnesota Press, 2013.
Hornaday, William T. "The Destruction of Our Birds and Mammals: A Report on the Results of an Inquiry." In *The Annual Report of the New York Zoological Society*, 2:77–107. New York, 1898.
———. "Diversions in Picturesque Game-Lands: Grand Bad-Lands and Mule Deer." *Scribner's Magazine*, July 1908, 1–17.
———. *The Extermination of the American Bison, with a Sketch of Its Discovery and Life History*. Washington DC: Government Printing Office, 1887. Reprinted as William Temple Hornaday, *The Extermination of the Bison*, ed. John Mack Faragher. Washington DC: Smithsonian Institution Scholarly Press, 2002.
———. *Our Vanishing Wildlife: Its Extermination and Preservation*. New York: Scribner's Sons, 1913.
———. *Wild Animal Interviews and Wild Opinions of Us*. New York: Scribner's, 1928.
Horowitz, Roger. *Putting Meat on the American Table: Taste, Technology, Transformation*. Baltimore MD: Johns Hopkins University Press, 2006.
Horowitz, Roger, Jeffrey Pilcher, and Sydney Watts. "Meat for the Multitudes: Market Culture in Paris, New York City, and Mexico City over the Long Nineteenth Century." *American Historical Review* 109 (2004): 1055–83.
Hoxie, Frederick E. *A Final Promise: The Campaign to Assimilate the Indians, 1880–1920*. Lincoln: University of Nebraska Press, 1984.
———. *Parading through History: The Making of the Crow Nation in America, 1805–1935*. Cambridge: Cambridge University Press, 1995.
Hungry-Wolf, Adolf, ed. *The Blackfoot Papers*. 4 vols. Browning MT: Blackfeet Heritage Center, 2006.
Igler, David. *Industrial Cowboys: Miller and Lux and the Transformation of the Far West, 1850–1920*. Berkeley: University of California Press, 2001.
Ingold, Tim. *The Perception of the Environment: Essays in Livelihood, Dwelling, and Skill*. New York: Routledge, 2000.
Isenberg, Andrew C. *The Destruction of the Bison: An Environmental History, 1750–1920*. Cambridge: Cambridge University Press, 2000.
———. "The Returns of the Bison." *Environmental History* 2, no. 2 (April 1997): 179–96.
Iverson, Peter. *When Indians Became Cowboys: Native Peoples and Cattle Ranching in the American West*. Norman: University of Oklahoma Press, 1994.
Jacoby, Karl. *Crimes against Nature: Squatters, Poachers, Thieves, and the Hidden History of American Conservation*. Berkeley: University of California Press, 2001.
Jameson, Sheilagh. *Ranches, Cowboys, and Characters: Birth of Alberta's Western Heritage*. Calgary: Glenbow Museum, 1987.

Jobson, Valerie. "The Blackfoot and the Rationing System." *Alberta History* 33, no. 4 (1985): 13–17.

Jones, Karen R. *Wolf Mountains: A History of Wolves along the Great Divide.* Calgary: University of Calgary Press, 2002.

Jones, Susan. "Becoming a Pest: Prairie Dog Ecology and the Human Economy in the Euroamerican West." *Environmental History* 4, no. 4 (1999): 531–52.

———. *Valuing Animals: Veterinarians and the Patients in Modern America.* Baltimore MD: Johns Hopkins University Press, 2003.

Jordan-Bychkov, Terry G. *North American Cattle-Ranching Frontiers: Origins, Diffusion, and Differentiation.* Albuquerque: University of New Mexico Press, 1993.

Kennedy, Margaret Anne. *The Whiskey Trade of the Northwestern Plains: A Multidisciplinary Study.* New York: P. Lang, 1997.

Keyser, James D. "The Warrior as Wolf: War Symbolism in Prehistoric Montana Rock Art." *American Indian Art Magazine* 32, no. 3 (2007): 62–69.

Khan, Yousuf, Manuel Mah, and Ziad Memish. "Brucellosis in Pregnant Women." *Clinical Infectious Diseases* 32, no. 8 (2001): 1172–77.

Klein, Kerwin Lee. *Frontiers of Historical Imagination: Narrating the European Conquest of Native America, 1890–1990.* Berkeley: University of California Press, 1997.

Knobloch, Frieda. *The Culture of Wilderness: Agriculture as Colonization in the American West.* Chapel Hill: University of North Carolina Press, 1996.

Koch, Hans Peter Gyllembourg. "Letters from the Musselshell, 1869–1870." Edited by Carl B. Cone. *Pacific Northwest Quarterly* 37, no. 4 (1946): 313–37.

———. *Splendid on a Large Scale.* Edited by Kim Scott. Helena MT: Bedrock Editions and Drumlummon Institute, 2009.

Ladner, Kiera L. "Women and Blackfoot Nationalism." *Journal of Canadian Studies* 35, no. 2 (2000): 35–60.

LaDow, Beth. *The Medicine Line: Life and Death on a North American Borderland.* New York: Routledge, 2001.

Langston, Nancy. *Forest Dreams, Forest Nightmares: The Paradox of Old Growth in the Inland West.* Seattle: University of Washington Press, 1996.

Lears, Jackson. *Fables of Abundance: A Cultural History of Advertising in America.* New York: Basic Books, 1994.

LeCain, Timothy. *Mass Destruction: The Men and Giant Mines That Wired America and Scarred the Planet.* New Brunswick NJ: Rutgers University Press, 2009.

Leonard, John William, and Frank Holmes, eds. *Who's Who in New York City and State.* Vol. 4. New York: L. R. Hamersly Company, 1909.

Lepley, John G. *Blackfoot Fur Trade on the Upper Missouri.* Missoula MT: Pictorial Histories, 2004.

Lewis, David Rich. *Neither Wolf nor Dog: American Indians, Environment, and Agrarian Change*. New York: Oxford University Press, 1994.

Lewis, Oscar. *The Effects of White Contact upon Blackfeet Culture with Special Reference to the Role of the Fur Trade*. Seattle: University of Washington Press, 1942.

Limerick, Patricia Nelson. *The Legacy of Conquest: The Unbroken Past of the American West*. New York: Norton, 1987.

Livingston, James. *Pragmatism, Feminism, and Democracy: Rethinking the Politics of American History*. New York: Routledge, 2001.

Long, Philip. *Seventy Years a Cowboy*. Billings MT: Cypress Books, 1976.

Lopez, Barry Holstun. *Of Wolves and Men*. New York: Scribner, 1978.

Lubetkin, M. John. *Jay Cook's Gamble: The Northern Pacific Railroad, the Sioux, and the Panic of 1873*. Norman: University of Oklahoma Press, 2006.

MacKenzie, John. *The Empire of Nature: Hunting, Conservation, and British Imperialism*. Manchester: Manchester University Press, 1997.

Malone, Michael P., and Richard B. Roeder, eds. *The Montana Past: An Anthology*. Missoula: University of Montana Press, 1969.

Mancall, Peter. *Deadly Medicine: Indians and Alcohol in Early America*. Ithaca NY: Cornell University Press, 1995.

Marshall, Robert. "The Problem of the Wilderness." *Scientific Monthly* 30, no. 2 (February 1930): 141–48.

McGowan, Dan. *Animals of the Canadian Rockies*. New York: Dodd, Mead and Company, 1936.

McManus, Sheila. *The Line Which Separates: Race, Gender, and the Making of the Alberta-Montana Borderlands*. Lincoln: University of Nebraska Press, 2005.

Mech, L. David, and Luigi Boitani, eds. *Wolves: Behavior, Ecology, and Conservation*. Chicago: University of Chicago Press, 2003.

Mercer, Asa Shinn. *The Banditti of the Plains*. Cheyenne WY: A. S. Mercer, 1894.

Merrill, Karen. *Public Lands and Political Meaning: Ranchers, Government, and the Property between Them*. Berkeley: University of California Press, 2002.

Milner, Clyde A., and Carol A. O'Connor. *As Big as the West: The Pioneer Life of Granville Stuart*. Oxford: Oxford University Press, 2008.

Mitchell, J., and K. Greer. "Predators." In *Game Management in Montana*, edited by Thomas Mussehl and F. W. Howell, 207–14. Helena: Montana Fish and Game Department, 1971.

Montana, Division of Wildlife. "Predator Management in Montana, Symposium Proceedings, January 8, 2000, Billings, Montana." Helena: Montana Fish, Wildlife and Parks, 2000.

Moore, John. "Cheyenne Work in the History of U.S. Capitalism." In *Native Americans and Wage Labor*, edited by Alice Littlefield and Martha Knack, 122–40. Norman: University of Oklahoma Press, 1996.
Mowat, Farley. *Never Cry Wolf: The Amazing True Story of Life among Arctic Wolves*. Toronto: McClelland and Stewart, 1963.
Nash, Roderick Frazier. *Wilderness and the American Mind*. 4th ed. New Haven CT: Yale University Press, 2001.
Nelson, George. "Trapping Wolves." *Canadian Cattlemen*, June 1944, 26.
New York Zoological Society. *Annual Reports of the New York Zoological Society*. New York: New York Zoological Society, 1897–1920.
Norton, Marcy. *Sacred Gifts, Profane Pleasures: A History of Tobacco and Chocolate in the Atlantic World*. Ithaca NY: Cornell University Press, 2008.
Nugent, David. "Property Relations, Production Relations, and Inequality: Anthropology, Political Economy, and the Blackfeet." *American Ethnologist* 20, no. 2 (May 1993): 336–62.
Osgood, Ernest Staples. *The Day of the Cattleman*. Minneapolis: University of Minnesota Press, 1929.
Overholser, Joel. *Fort Benton: World's Innermost Port*. Fort Benton MT: J. Overholser, 1987.
Pallister, H. G. "Smoke from the Branding Fire." *Alberta Beef*, August 1997, 40–43.
Petersen, David. *Heartsblood: Hunting, Spirituality, and Wildness in America*. Washington DC: Island Press, 2000.
Pilcher, Jeffrey. *¡Que vivan los tamales! Food and the Making of Mexican Identity*. Albuquerque: University of New Mexico Press, 1998.
Prucha, Francis Paul. *American Indian Policy in Crisis: Christian Reformers and the Indian, 1865–1900*. Norman: University of Oklahoma Press, 1976.
———. *The Great Father: The United States Government and the American Indians*. Lincoln: University of Nebraska Press, 1984.
Rawson, Timothy. *Changing Tracks: Predators and Politics in Mt. McKinley National Park*. Fairbanks: University of Alaska Press, 2001.
Ray, Arthur J. *The Canadian Fur Trade in the Industrial Age*. Toronto: University of Toronto Press, 1990.
Ripley, William Zebina. *The Races of Europe: A Sociological Study*. New York: D. Appleton and Company, 1899.
Robinson, Michael J. *Predatory Bureaucracy: The Extermination of Wolves and the Transformation of the West*. Boulder: University Press of Colorado, 2005.
Rorabaugh, W. J. *The Alcoholic Republic: An American Tradition*. Oxford: Oxford University Press, 1979.
Rorty, Richard. *Contingency, Irony and Solidarity*. Cambridge: Cambridge University Press, 1989.

Rosier, Paul C. *Rebirth of the Blackfeet Nation, 1912–1954*. Lincoln: University of Nebraska Press, 2001.
Rothfels, Nigel. *Savages and Beasts: The Birth of the Modern Zoo*. Baltimore MD: Johns Hopkins University Press, 2002.
Ruppel, Kristin. *Unearthing Indian Land: Living with the Legacies of Allotment*. Tucson: University of Arizona Press, 2008.
Russell, Charles M. "Whiskey." In *Montana Margins: A State Anthology*, edited by Joseph Kinsey Howard, 378–81. New Haven CT: Yale University Press, 1946.
Samek, Hana. *The Blackfoot Confederacy, 1880–1920: A Comparative Study of Canadian and U.S. Indian Policy*. 1st ed. Albuquerque: University of New Mexico Press, 1987.
Schultz, James Willard. *Many Strange Characters: Montana Frontier Tales*. Norman: University of Oklahoma Press, 1982.
———. *The Starving Blackfeet Indians*. Los Angeles: National Association to Help the Indian, 1921.
Sharp, Paul F. "The Northern Great Plains: A Study in Canadian-American Regionalism." *Mississippi Valley Historical Review* 39, no. 1 (1952): 61–76.
———. *Whoop-Up Country: The Canadian-American West, 1865–1885*. Minneapolis: University of Minnesota Press, 1955.
Simmel, Georg. *The Philosophy of Money* [1900]. Translated by David Frisby and Tom Bottomore. London: Routledge, 1990.
Slotkin, Richard. *The Fatal Environment: The Myth of the Frontier in the Age of Industrialization, 1800–1890*. New York: Atheneum, 1985.
———. *Gunfighter Nation: The Myth of the Frontier in Twentieth-Century America*. New York: Atheneum, 1992.
Smith, Bruce L. *Where Elk Roam: Conservation and Politics of Our National Elk Herd*. Guilford: Globe Pequot Press, 2012.
Smith, Bruce L., Eric Cole, and David Dobkin. *Imperfect Pasture: A Century of Change at the National Elk Refuge in Jackson Hole, Wyoming*. Jackson: Grand Teton Natural History Association, 2004.
Smith, Douglas W. *Decade of the Wolf: Returning the Wild to Yellowstone*. Guilford CT: Lyons Press, 2005.
Smith, Linda Tuhiwai. *Decolonizing Methodologies: Research and Indigenous Peoples*. London: Zed Books, 1999.
Spence, Mark David. *Dispossessing the Wilderness: Indian Removal and the Making of the National Parks*. New York: Oxford University Press, 1999.
Spivak, Gayatri Chakravorty. "Scattered Speculations on the Question of Value." *Diacritics* 15, no. 4 (1985): 73–93.
State of Montana. *Laws, Resolutions, and Memorials of the State of Montana*. Helena: various publishers, 1890–1930.

Steele, Jordan, et al. "Wolf Predation Impacts on Livestock Production: Direct Effects, Indirect Effects, and Implications for Compensation Ratios." *Rangeland Ecology and Management* 66, no. 5 (September 2013): 539–44.

Stegner, Wallace Earle. *The Selected Letters of Wallace Stegner*. Emeryville CA: Shoemaker & Hoard, 2007.

———. *Wolf Willow A History, a Story, and a Memory of the Last Plains Frontier*. New York: Viking Press, 1962.

Steinhart, Edward. *Black Poachers, White Hunters: A Social History of Hunting in Colonial Kenya*. Athens: Ohio University Press, 2006.

Stern, Alexandra. *Eugenic Nation: Faults and Frontiers of Better Breeding in Modern America*. Berkeley: University of California Press, 2005.

Sutter, Paul. *Driven Wild: How the Fight against Automobiles Launched the Modern Wilderness Movement*. Seattle: University of Washington Press, 2005.

Taussig, Michael. *Mimesis and Alterity: A Particular History of the Senses*. New York: Routledge, 1993.

Territory of Montana. *Laws, Resolutions, and Memorials of the Territory of Montana*. Helena: George Boos, 1883–89.

Thomas, David Hurst. *A Blackfoot Source Book*. New York: Garland Publishing, 1986.

Thompson, E. P. *Whigs and Hunters: The Origin of the Black Act*. 1st ed. New York: Pantheon Books, 1975.

Toole, Allan. "Wolf Hunt Summer in Northeastern Montana, 1911." *Montana: The Magazine of Western History* 37, no. 3 (1987): 66–70.

Trosper, Ronald Lloyd. "The Economic Impact of Allotment Policy on the Flathead Indian Reservation." PhD diss., Harvard University, 1974.

Tuan, Yi-Fu. *Dominance and Affection: The Making of Pets*. New Haven CT: Yale University Press, 1984.

Turner, Frederick Jackson. "The Significance of the Frontier in American History." In *Rereading Frederick Jackson Turner*, edited by John Mack Faragher, 31–60. New Haven CT: Yale University Press, 1999.

Uhlenbeck, C. C. *An English-Blackfoot Vocabulary Based on Material from the Southern Piegans*. Amsterdam: Koninklijke Akademie van Wetenshappen te Amsterdam, 1930.

United States Bureau of Indian Affairs. *Reports of the Inspection of the Field Jurisdictions of the Office of Indian Affairs, 1873–1900*. Washington DC: National Archives Microfilm Publications, M1070.

United States Department of Fish and Wildlife. *The Reintroduction of Gray Wolves to Yellowstone National Park and Central Idaho: Final Environmental Impact Statement*. Washington DC: Government Printing Office, 1994.

Van Nuys, Frank. *Americanizing the West: Race, Immigrants, and Citizenship, 1890–1930*. Lawrence: University Press of Kansas, 2002.

Vialles, Noelie. *Animal to Edible*. Translated by J. A. Underwood. Cambridge: Cambridge University Press, 1994.

Voggesser, Garrit. "Of Woods, Wilderness, and Water: Negotiating Natural Resources on the Blackfeet, Flathead, and Fort Peck Reservations, 1885–1945." PhD diss., University of Oklahoma, 2004.

Walker, Brett L. *The Lost Wolves of Japan*. Seattle: University of Washington Press, 2005.

Warren, Louis. *The Hunter's Game: Poachers and Conservationists in Twentieth-Century America*. New Haven CT: Yale University Press, 1999.

Welch, James. *Fools Crow*. New York: Penguin, 1986.

Welch, James, and Paul Stekler. *Killing Custer: The Battle of Little Bighorn and the Fate of the Plains Indians*. New York: Penguin, 1995.

Wessel, Thomas. "Agriculture on the Reservations: The Case of the Blackfeet, 1885–1935." *Journal of the West* 18 (October 1979): 17–24.

West, Elliot. *The Contested Plains: Indians, Goldseekers, and the Rush to Colorado*. Lawrence: University Press of Kansas, 1998.

———. *The Saloon on the Rocky Mountain Mining Frontier*. Lincoln: University of Nebraska Press, 1979.

Whaley, Gray. *Oregon and the Collapse of Illahee: U.S. Empire and the Collapse of an Indigenous World*. Chapel Hill: University of North Carolina Press, 2010.

Whealdon, Bon I. *"I Will Be Meat for My Salish": The Montana Writers Project and the Buffalo of the Flathead Indian Reservation*. Pablo MT: Salish-Kootenai College, 2001.

Wheeler, Burton K., with Paul F. Healy. *Yankee from the West*. New York: Doubleday, 1962.

White, G. Edward. *The Eastern Establishment and the Western Experience: The West of Frederic Remington, Theodore Roosevelt, and Owen Wister*. Yale Publications in American Studies 14. New Haven CT: Yale University Press, 1968.

White, Richard. *The Organic Machine: The Remaking of the Columbia River*. New York: Hill & Wang, 1995.

Whittlesey, Lee. "Cows All Over the Place: The Historic Setting for the Transmission of Brucellosis to Yellowstone Bison by Domestic Cattle." *Annals of Wyoming* 66 (Winter 1994–95): 42–57.

Willerslev, Rane. *Soul Hunters: Hunting, Animism, and Personhood among the Siberian Yukaghir*. Berkeley: University of California Press, 2007.

Wilson, Robert. *Seeking Refuge: Birds and Landscapes of the Pacific Flyway*. Seattle: University of Washington Press, 2012.

Wissler, Clark. "Ceremonial Bundles of the Blackfoot Indians." *Anthropological Papers of the Museum of Natural History* 7, no. 2 (1912).

———. "Material Culture of the Blackfoot Indians." *Anthropological Papers of the Museum of Natural History* 7, no. 1 (1910).

———. "The Social Life of the Blackfoot Indians." *Anthropological Papers of the Museum of Natural History* 7, no. 1 (1910).

———. "Societies and Dance Associations of the Blackfoot Indians." *Anthropological Papers of the Museum of Natural History* 11, no. 4 (1916).

Wissler, Clark, and Alice Beck Kehoe. *Amskapi Pikuni: The Blackfeet People*. Albany: State University of New York Press, 2012.

Woehlke, Walter V. "Hope for the Blackfeet." *Sunset Magazine*, December 1923, 9–11.

Wolch, Jennifer R., and Jody Emel, eds. *Animal Geographies: Place, Politics, and Identity in the Nature-Culture Borderlands*. London: Verso, 1998.

Wolf, Eric. *Europe and the People without History*. Berkeley: University of California Press, 1982.

Wolfe, Cary. *Animal Rites: American Culture, the Discourse of Species, and Posthumanist Theory*. Chicago: University of Chicago Press, 2003.

Worster, Donald. *Nature's Economy: A History of Ecological Ideas*. 2nd ed. Cambridge: University of Cambridge Press, 1994.

———. *Rivers of Empire: Water, Aridity, and the Growth of the American West*. Oxford: Oxford University Press, 1986.

Young, Stanley Paul. *The Wolf in North American History, with Fifty-Three Illustrations*. Caldwell ID: Caxton Printers, 1946.

Zontek, Ken. *Buffalo Nation: American Indian Efforts to Restore the Bison*. Lincoln: University of Nebraska Press, 2007.

Index

Adams, Charles, 158n74
agriculture, xvi, xix–xxi, 15, 23, 42, 54–57, 69–74, 90–97; as alternative to industrial capitalism, 71–74, 91–97; and bounty hunting, 23; and Canadian politics, 42; and colonial notions of predation and production, 21, 70–71; displacing the environmental effects of, 100–101, 128–29; and dispossessions of Native land, 15, 54–57, 69–70, 90, 142n13; and dryland farming, 42, 69; history of, 73–74, 90; as scientific enterprise, xvi, 128–29; and subordination of Native labor, 4, 54–57, 69–70, 90. *See also* cattle; farming; food
Alberta, xvi, xviii, xix, xxii, 1, 5; conservation in, 100, 104, 106, 127, 130; dispossession of indigenous communities in, 1, 5, 21, 53, 65, 70, 73, 89; provincial government of, 34–35, 40–42; wolfing in, 25–27, 29, 32–33, 39–44

Allard, Charles, 99, 105, 107
allotment, xviii, xix, xxi, 12–13 69–73, 83–98, 105–7; of the Blackfeet Reservation, 69–73, 79, 83–89, 97–98; and creation of National Bison Range, 105–7; of the Flathead Reservation, 105; history of, 71, 89–90
American Bison Society, 100–132; and fundraising efforts, 101–13; and narratives of predation, 103–7, 114–26
American Fur Company, 7–9, 16
American Society of Mammalogists, 103, 128–29, 158n74
Animal Damage Control Act, 102–3, 129–30
animals, xvi–xxi, 8, 10–12, 25–27, 30–37, 44–66, 70–75, 80–85, 103–9, 112–13, 127, 133–38, 139n3, 140n7, 141n11, 142n15, 143n20, 144n19; and agriculture, xviii, 48, 51, 53–55, 73, 75, 77, 80–81, 85, 123, 139n3; as attractions, 99, 103–4, 107–13, 120, 123, 133–38; as bodies, xvi–xvii,

175

animals (cont.)
  xxii, 25–27, 30, 33–37, 48–49, 60; communication with, 4, 10, 136, 140n7; and disease, 16, 59; as food, xviii, xx, xxii, 7, 24–27, 30, 48–50, 53, 55, 58, 60–64, 75, 80–81, 98, 100, 108, 113, 129–32, 135–38, 139n3; and historical agency, xxii, 33, 120, 127–29, 140n7, 142n15, 143n20, 144n19; and kinship, xvii–xviii, 10–12, 25–27, 33, 49, 65, 70, 74, 92, 141n11, 144n19, 145n21; as laborers, xvi–xix, 8, 23–27, 31–32, 44–45, 48–49, 65–66, 70, 74, 83, 92, 98, 110, 133–38, 139n3, 142n15; as pets, xxiii, 112–13, 135–37; as property, xvii, 31–33, 35–38, 42, 51, 53, 56–57, 60–66, 74–75, 85, 92, 98, 102–3, 106–7, 109, 127, 133–38; as scientific subjects, xvi, 128–30, 133–35, 138, 158n74. *See also individual animals*; labor (nonhuman)
animism, 3, 11, 65, 140n7
Anoka County (Minnesota), 113
Arapaho people, 51
Assiniboine people, 7, 17, 20

Bad Bob (bison), 113
Badger Creek Agency, 49, 53, 58, 61
Bailey, Vernon, 35
Baker, Eugene, 3, 52–53
Baker, Isaac G., 6, 8, 14–15, 17, 56–57
Banff (Alberta), 109
Banff National Park (Alberta), 99, 101, 131
Bantjes, Rod, 72, 142n12
barbarism, xvii, 49, 58–59, 65–66, 138
Barret, Maurice, 30, 135
Bar U Ranch, 32

Basques, 24, 100, 119–26, 130
Bastien, Betty, xix, 11–12
Battle of Poker Flats, 53
bears, 33–34, 44, 122, 138
beavers, 7
Benedict, Samuel, 55
Benton, Thomas Hart, 8
Billings (Montana), 115, 131
Binnema, Ted, 132
bison, xvii–xviii, xx, xxii, 3, 5–8, 12, 15, 18, 20–21, 26, 28–29, 47–51, 53, 55, 60, 65–66, 84, 99, 100–132, 145n31, 146n39, 150n21, 154n2; Blackfoot relationships with, 47–51, 53, 55, 60, 65–66; conservation of, 100–132. *See also* American Bison Society; animals; fur trade; Hornaday, William; National Bison Range
Blackfeet Reservation, xix, 9, 15, 47, 74–82; and boundary fence 76–77; and cattle trespass, 76–79; and class divisions, 74–75, 79–82; and grazing taxes 77–79
Blackfeet Stock Protective Association, 81–82
Blackfeet Tribal Business Council, 86, 89, 91
Blackfoot people xvii–xviii, 3–13, 47–98, 141n8; and cultural ceremonies, 9–10, 13, 93, 153n44; and hunting practices, xvii, xxi, 7–8, 16–17, 49–50; and language, 72, 90, 92, 97, 151n2; and nonhuman kinship, xvii–xviii, 3–5, 10–13, 70–71; and terminology, 141n8
Blackfoot Reserve, 21
Black Panther Oil Company, 117
Blacktail Mountains, 134
Blood Reserve, 21

boarding schools, xviii, 64
bounty programs, xv, xix–xx, 23–46, 102, 120, 133–35; in Alberta vs. Montana, 40–42; economic conditions of, 27–29, 33, 36–45, 133–35; fraud, 38–39; laws concerning, 33–35; numbers killed by, 33–34. *See also* wolfing
Bozeman (Montana), xvi, 52
Bozeman National Bank, xvi
Brackett, Oscar, 19
bread, xviii, 64
Broadwater, E. J., 38
Bronx Zoo, 104, 108–9
Brower, Jennifer, 131
Brown, Walter, 38
Browning (Montana), 49, 58, 64, 75, 78, 80, 86–89, 91
brucellosis, 16, 154n2
butchery 16, 49–50, 57–66

Calgary (Alberta), 4
Cameron, Evelyn, 30, 135–37
Cameron, Ewen, 30, 32, 38, 135–37
Campbell, Fred, 73, 91–92
Canada, xviii, xx; agricultural politics in, 32, 42, 97; cattle in, 33, 39–40, 42; colonial process in 8, 12, 51, 53, 55, 91, 115, 122, 141n8; and conservation, 100, 104, 106, 127–28, 131; and the whiskey trade, 4–7, 86; wolfers in, 18, 28, 32, 29
capitalism, xvi, xx; and animal bodies, 24, 33, 37–39, 56, 64, 74–75, 83, 98; as colonial process, 8, 12, 45, 47–53, 56, 64–66, 70–72, 75, 77, 83, 93; and conservation, 102–3, 109–10, 114, 118, 128–30, 132, 134, 138; meat and, xvi, 45, 47, 49, 51, 53, 55–56,
64–66, 70; selfhood and, 70–72, 75, 77, 83, 90; and the whiskey trade, 3, 8, 13–14; and wolfing, 23–24, 33. *See also* class; colonialism; labor (human); labor (nonhuman)
Carhart, Arthur, 103, 126–29
Cat Creek oil field (Montana), 114, 117
Celilo Falls (Oregon), 139n3
Chang, David, 90
Cheyenne people, 10, 51
Chicago (Illinois), 7, 15–16, 37, 58, 62, 69, 139n2
Choteau County (Montana), 56
Churchill, Clarence, 76–81, 83, 86
civilization, xv, 135, 138; and dissociation of predation from production, 21, 48, 50, 53, 55, 58–59, 62, 65–66, 104, 109, 120, 123, 126–28, 135, 138
Clark, Badger, 115
class, xv, xx, 30: and Blackfeet social divisions, 74–75, 77, 82; and the conservation movement, 109–10, 119, 122; and wolfing, 15, 24, 30, 41. *See also* capitalism; labor (human); labor (nonhuman)
Clendinnen, George, xv, 19
Cobell, Eloise, 90
Cody (Wyoming), xiii, 7
coffee, 50, 54
Coleman, Jon, 31, 140n5
Collier, John, 91
colonialism, xvi; alternatives to, xvii, xix, 42, 90, 98; and assimilation programs, xviii, 48, 53, 65–66, 122; and the conservation movement, 100–103, 108, 114, 127, 134–8; and decolonizing scholarship, 141n11; disavowed histories of, 3–5, 45, 98–99, 108, 127; and the dissociation

colonialism (*cont.*)
   of predation from production, xvi, xviii, xix–xx, 4, 13, 21–26, 31, 37, 44, 48–49, 53, 62–67, 83, 98–102, 108, 115, 129, 131, 134, 138; and ecological transformations, xvi, 22–26; food and, 48–49, 53, 62, 64–67, 83–84; and gender anxieties, 112; and historical agency, 3, 132; and ideas about legitimate work, xx, xxiii, 23–25, 28, 48–49, 53, 62, 64–67, 82, 134, 142n15; and inversion of violence, 31; language of, xvii; and ontology of "natural alliance" (Bastien), 11–12, 144n21; and privatization, xvi, 21, 90–97, 102, 143n19; and relations with animals, xvii, xix, xxiii, 3, 33, 140n5; resistance (indigenous) to, xviii, xix, 5, 39, 90–96; resistance (non-indigenous) to, 15, 40, 42, 145n30; resistance (non-human) to, 28; and the scientific establishment, 11, 128–29, 138; and selfhood, xvii, xix, 4, 9–13, 65, 71–72, 144n19; through stomachs, 48; and whiskey, 3–5, 13
Columbia River, 139n3
Coney Island (New York), 135
Conrad family, 14; and Alice Conrad, 107; and Joseph Conrad, 107; and bison herd, 107, 111. *See also* American Bison Society; National Bison Range
conservation, xxi, 98–132, 138; and dispossession of indigenous land, 101–13; and narratives of predation, 103–7, 114–26
Cooper, Walter, 14

coyotes, xv, 11, 20, 27–28, 30, 33–34, 36, 38, 40–41, 104, 119, 126–27, 130
Creek people, 90
Cree people, 7, 10, 17, 38
crocodiles, 138
Cronon, William, 139n2
Crow people, 20, 51
Crow Reservation, 78, 81
Curly Bear, 9
Curnow, Edward, 28
Custer, George, 31
Cut Bank (Montana), 88
Cypress Hills (Saskatchewan), 31
Cypress Hills massacre, 20, 112

Dare, J. Z., 76, 80
Davis, Sam, 36
Dawes, Henry, 82
deer, 55
Delmonico's Restaurant, 132
Dempsey, Hugh, 3
Dempsey, Jack, 92
Denver Public Library, xxiii
Derrida, Jacques, 140n7
DeVoto, Bernard, xxi
Division of Predator Animal and Rodent Control. *See* U.S. Bureau of Biological Survey
Dixon (Montana), 105
Dixon, Joseph, 105
Duerr, Hans Peter, 133
dupuyer, 50
Duvall, David, 9

Edmonton (Alberta), xiii, 7, 39, 41
elk, xxii 26, 55, 133–34
Elk-horn, 47
Elliot, Howard, 112
Ellis, Vivian, 31

Elofson, Warren, 39
Elrod, Morton J., 105
Emel, Jody, 135
Evans, John, 20
Ewers, John, 73, 83–84

Farm Bureau, 74
farming. *See* agriculture
Farr, William, 61
Farwell, Abel, 20
Felton, W. T., 117
Fiege, Mark, 72
firearms, 6, 8, 15, 17–21, 29, 31, 49–52, 109, 113, 115, 118
Fitzgerald, Deborah, 72
Flathead Reservation, 99, 105–7
Flores, Dan, 2, 132, 145n31
flour, 50, 54, 56, 64, 91, 94
Flowerree Cattle Company (Montana), 76
Flowerree, Dan, 76
food, xvi–xviii, xix; on the Blackfeet Reservation, 47, 49–58, 60, 63–66, 70–74, 83–85, 93, 95–97; canned, 6; and tastes for the "frontier," 99, 132; for wolves, 18, 28, 30, 134. *See also* meat
Fort Benton (Montana), 6–8, 13–16, 20–21, 51, 56, 107
Fort Ellis (Montana), 52
Fort Macleod (Alberta), 40–41
Fort Shaw (Montana), 13–14, 52
Fort Union (Montana), 7–8
Fort Whoop-Up (Alberta), 6, 20, 86
foxes, 11, 20, 27
fraud, 37–39
fur trade, xv, xix–xxi, 5–8, 12, 15–17, 19–23, 26–28, 38, 45, 50–51, 111, 146n39

gender: and Blackfoot labor, 16–17, 32, 48–49, 56–57, 59–64, 66; and masculine performance within the conservation movement, 115–19, 122
Gibbons, Tom, 92
Glacier National Park (Montana), 84, 99, 101, 112
Goetzmann, William, 19
Graham, Alistair, 138
Grant, Madison, 101, 119–26
Grant, Ulysses, 54
Graybill, Andrew, 53
Great Northern Railroad, 80, 88, 92, 107, 112
Greenwald, Emily, 90
Grinnell, George Bird, 11, 123
Grinnell, Joseph, 158n74
guns. *See* firearms

Hagenbeck, Carl, 108
Hall, E. Raymond, 158n74
Hamilton, Alfred B., 6, 86
Hamilton, Robert, 72, 84–91
Haraway, Donna, 136, 159n6
Hardwick, Thomas, 20, 112
Healy, John J., 6, 20
Heart Butte (Montana), 91
Helena (Montana), 7, 34
Hill, James J., 107, 112–13
history, xvii; and animals, xxiii, 25, 140n5, 140n7; Blackfeet concepts of, 47–48, 65; and colonialism, xix, 25, 45, 47–48, 61, 65, 72–73, 84, 90, 100, 145n21; and the "frontier," 127; of immigration, 124–25; language and, 141n7; and narratives of predation, 108, 110–11; and nonhuman agency, 10, 21, 25, 45, 100; of wildlife conservation, 103–4, 108, 110, 131–32, 134–35

Hornaday, William, 100–132, 146n39
horses, 2, 14, 16, 20, 30, 33, 49–52, 74, 77, 81, 86, 92, 113
Howard, C. H., 56, 59
Howell, A. Brazier, 103
Hoxie, Frederick, 71
Huckvale, Walter, 33, 42
Hudson River, 118
Hudson's Bay Company, 7, 16
Huffman, Laton Alton, 114
Hughes, Gerald, 29
hunting, xvii–xviii, xix–xxii; Blackfeet practices of, 5–7, 15–17, 20–21; valued as illegitimate labor, 44–45, 47–55, 57–58, 60, 64–66, 70, 72, 75, 82, 84, 95, 100, 102, 108–9, 113, 115, 120, 123–24, 131, 133

Immigration Exclusion Act of 1924, 123, 130
Indian police, 57–58
Indian Reorganization Act, 89
Ingold, Timothy, 144n21
Irvin, Louis S., 89
Isenberg, Andrew, 130–32, 145n31, 146n39

Jacoby, Karl, 109
Johns Hopkins University, 103
Jones, Karen, 140n5
Judith Basin (Montana), 29, 108

Kainah. *See* Blackfoot people
Kalispell (Montana), 107
Katmai National Park (Alaska), 138
Kennard, Fred, 109, 111
Kennedy, Margaret, 4
kinship, xvii–xviii, xix, 4–5, 145n21; and allotment, 94–95; Blackfeet understandings of, 10–13, 49, 65, 70–72, 75, 90, 92, 98
Knobloch, Frieda, 142n12
Koch, Peter, xv–xvi, 19, 25–27

labor (human), xv–xviii, xix–xxi, 4, 15–17, 19, 23–28, 31–32, 35, 40, 44–45, 47–67, 70–75, 79, 82–83, 86, 90, 93, 99–105, 107, 110–19, 134, 138
labor (nonhuman), xvii–xviii, xix, xxi, 4, 15, 25–26, 31–32, 35, 44, 65–66, 71–75, 79, 83, 89, 90, 97, 98, 139n4
labor theory of value, xx, 25, 32, 40, 44, 65; and value-theory of labor, 142n15
Lakota people, 10, 51
Lane, George, 32, 42
Lane, Harry, 87–88, 120
Lears, Jackson, 32–33, 72
LeCain, Timothy, 143n18
Leopold, Aldo, 128
Lewis, David Richard, 142n13
Lewiston (Montana), 114
Lipps, Oscar, 85, 95
livestock industry, xvi, 23; and representations of production, 24–25; and public subsidies, 35–37
Livingston, James, 72
Long, T. B., 29

Marias Massacre, 51–53
Marias River (Montana), 1, 52–53
Marshall, Bob, 2
Marshall, Duncan, 42
Massachusetts, 111
McFatridge, Arthur, 76, 80–82, 84–85, 88–89, 93
McGowan, Dan, 18
McKnight, Joseph, 13–14

180 INDEX

McLeod, James, 3
Meagher, George, 14–15
meat, xvi, xix, xxii, 16–17, 49–68
Mech, David, 146n43
Medicine Hat (Alberta), 41
Merrill, Karen, 147n1
Miles City (Montana), 115, 124
Milk River, 1
Minneapolis, 99, 111–13
Minneapolis Commercial Club, 111
Minneapolis Tribune, 99, 104, 113
Mississippi, xv
Missoula (Montana), 105, 119
Missouri River, xv, 1, 6–7, 19, 51, 54–55, 119, 124
Montana: state of, 1, 35–36, 38, 77; territory of Montana, 27, 33–34
Montana Mutual Oil Syndicate, 117
Montana State University, xvi
Montana Stock Grower's Journal, 114
Montana Stock Growers' Association, 27–28
Monteath, James, 94
Montreal, 7
Moore, John, 51
Mountain Chief, 52, 87
mountain goats, 109
mountain lions, 33–34, 44, 130
Mowat, Farley, 138
Murie, Olaus, 128
Musselshell River, xv, 19

Nakoda people, 39
National Automobile, 113
National Bison Range (Montana), xxii, 99, 102–14, 132
National Elk Refuge (Wyoming), xxii
Nelson, E. W., 120
Nelson, George, 28

neurasthenia, 112
Newfoundland, 109
New Orleans, 99
New York, 7, 99, 101, 111, 117
New York Zoological Society, xxiii, 110
Newberry Library, xxiii
Niitsitapi. *See* Blackfoot people
Nonpartisan League, 74, 89
Northern Pacific Railroad, 112, 132
North Oaks (Minnesota), 107, 112–13
Northwest Fur Company, 8
North-West Mounted Police, 3, 20, 53
Northwest Territories, 34
Nugent, David, 51

oil, 89, 96, 114–19, 146n49
Oklahoma, 104, 122, 156n49
Oldman River, 6
Oregon, 120–21
Osborn, Henry Fairfield, 124–25
Ottawa, 39

Pablo, Michel, 99, 105–7
painting, 1–2, 32. *See also* Russell, Charles M.
Pallister, H. G., 29
Pawnee people, 9
pelts. *See* fur trade
pemmican, 50
Petersen, David, 138
pets, xxiii, 135–37. *See also* animals
Pettigrew, Richard, 112
peyote, 9–10
Philadelphia (Pennsylvania), 117
photography, 30, 61, 91, 114–17, 123–24, 135–37. *See also* Cameron, Evelyn; Huffman, Laton Alton
Piegan Farming and Livestock Association, 70–75, 89–98

Piegan Reserve, 21
Pikuni people. *See* Blackfoot people
Pincher Creek, 41
Pine Ridge Reservation, 81
pisskans, 49
plains: descriptions of, 1–3
Power, Thomas C., 6, 15, 17–18, 28, 55
prairie dogs, 33–34
predator-prey relationships: and historical contingencies of, xvii, 29, 102, 135, 138, 139n4. *See also* animals; capitalism; colonialism; labor (human); labor (nonhuman)

race: and Blackfoot understandings, 10–13; distinctions between full blood and mixed blood, 84–85; and predispositions toward cruelty, 120–26
ranching. *See* agriculture; livestock industry
Rawson, Timothy, 140n5
Redington, Paul, 103
Regina (Saskatchewan), 39
Remington, Frederic, 2, 115, 120–23
Riviere, H. A., 26–27, 39
robes. *See* fur trade
Robinson, Michael, 140n5
Rocky Boy's Reservation, 38
Roosevelt, Theodore, 101, 109–10, 115–17, 122
Rosebud Reservation, 81
Rosier, Paul, 73, 84
Rowe, Charlie, 19
Russell, Charles M., 1–5, 20, 32, 115, 123

Salish-Kootenai people, xviii, 105; and contributions toward bison conservation, 105–9
Samek, Hannah, 150n20

San Angelo (Texas), 118
San Francisco, 99
Saskatchewan River, 6
sausage, 50
Schreyvogel, Charles, 2
Schultz, James Willard, 96
science: and capitalism, xvi; and colonialism, 3; and emerging field of ecology, 128–31; and natural history paradigm, 108, 120, 128, 130, 157n56; and predator control, 103, 118, 126, 131; and wildlife management, xxi–xxii, 99–102, 138, 143n21
Scott, H. L., 95, 97
Scribner's, 124
selfhood: Blackfoot understandings of, 10–12, 70–73, 84, 90–97; and capitalism, 64–65, 70–73, 83–84, 89–92; and nostalgia for the frontier, 1, 122; and wolfing, xv–xvii, 30–32, 144n19
Seymour, Edmund, 101, 118–19
Sharp, Paul, 4, 20
Shelby (Montana), 92
Sheridan, Philip, 14, 52
Siksika people. *See* Blackfoot people
Simpson, James, 109
slaughterhouses, xxi, 48–49, 53–66, 71–74, 81, 139n3
Slotkin, Richard, 31, 157n61
Smith River, 31
Smith, Linda Tuhiwai, 141n11
Smithsonian Institution, 108, 132
Snowdrift (wolf), 29
South Dakota, 112
St. Louis (Missouri), xv, 1–2, 7–8, 99
St. Marys River, 6
St. Paul (Minnesota), 7–8, 107, 111–13
steamboat, 3
Steell, George, 75–76

Stegner, Wallace, 2, 69
strychnine, 17–20, 26–27
Stuart, Granville, 81
sugar, 50, 54
Sun River, 1
Sun River Stock and Land Company, 76
surplus killing, 134
Sweetgrass Hills massacre, 20

Tacoma (Washington), 118
Taussig, Michael, 11
tea, 9, 50
Teton River, 1
Texas, 80, 118
Treadwell, Timothy, 138
Treaty Seven, 3, 53
Troy, T. J., 38
Turner, Frederick Jackson, 118
Two Medicine irrigation project (Blackfeet Reservation), 86–87
Tyrannosaurus rex, 124

University of Montana, 105
U.S. Army, 6, 8, 13–14, 20, 51–55
U.S. Bureau of Biological Survey, xvii, 27, 35, 45; and Division of Predator Animal and Rodent Control, 103, 126–30
U.S. Bureau of Reclamation, 81
U.S. National Archives, xxiii
U.S. National Forest Service, 127
U.S. Office of Indian Affairs, xxi, xvii, 6, 47–48, 53–55, 70–98
U.S. Senate Committee on Indian Affairs, 89, 105

Vale, Charley, 29
Van Nuys, Frank, 125

Vialles, Noelie, 139n2
Volstead Act, 2

Wainwright Buffalo National Park (Alberta), xxii, 99–100, 102, 104–7, 126, 131
Walker, Brett, 140n5
Wallace, William, Jr., 76
Washington Post, 104, 122
Wessel, Thomas, 73
Western Stock Growers' Association, 28, 33–34, 39–42
Whaley, Gray, 145n30
Wheeler, Burton K., 89, 91
whiskey: as "dissolute theory of conquest," 1–3; environmental effects of, 5, 17; and historical agency, 3–6; and intoxicating effects, 1–3, 9–14; social effects of, 4–8, 15–21; traders, xv, xx, 6–8, 15–21, 34, 50, 55, 84, 86, 107
White, Richard, 139n3
Wibaux, Pierre, 28
Wichita Mountains (Oklahoma), 104, 122
Wild Animal Interviews (Hornaday), 120
Willerslev, Rane, 11, 141n7, 144n19
Williams College, 118
Wilson, Robert, 143n21
Winnipeg (Manitoba), 39
winter: of 1906–1907, 79–80; of 1919–1920, 69, 79
Wissler, Clark, 9–10
wolfing, xv; by denning, 30–31; with firearms, 29–30; for pelts, xv–xvi, 6–8, 17–20, 23–33; for bounty payments, xx–xxi, 33–45
"The Wolf-Man" (Grinnell), 11

wolves: artistic representations of, 2–3, 21–22, 135–38; eradication of, xv–xviii, xix–xxi, 23–45, 100, 102–4, 109, 119–20, 125–30, 135, 140n5; estimated historical populations of, 18, 146n39; as historical agents, 4, 5, 17–18, 21–22, 136; killing livestock, 27–28, 133–34; kinship with humans, xv–xvii, xxi, 11, 25–26, 49, 57, 134–38; as means of production, xv, 7, 17–22, 33–45, 135; as pets, 135–37; reintroduction to Northern Rockies, xxii, 133–34

Wren, George, 95

Wyoming, 118–19

Yellowstone National Park (Wyoming), 104, 133–34, 154n2

Yellowstone River, 7, 53

Young, John, 54–57

Young, Stanley, 103, 126–30, 146n43

www.ingramcontent.com/pod-product-compliance
Lightning Source LLC
Chambersburg PA
CBHW022101160426
43198CB00008B/304